SpringerBriefs in Geography

SpringerBriefs in Geography presents concise summaries of cutting-edge research and practical applications across the fields of physical, environmental and human geography. It publishes compact refereed monographs under the editorial supervision of an international advisory board with the aim to publish 8 to 12 weeks after acceptance. Volumes are compact, 50 to 125 pages, with a clear focus. The series covers a range of content from professional to academic such as: timely reports of state-of-the art analytical techniques, bridges between new research results, snapshots of hot and/or emerging topics, elaborated thesis, literature reviews, and in-depth case studies.

The scope of the series spans the entire field of geography, with a view to significantly advance research. The character of the series is international and multidisciplinary and will include research areas such as: GIS/cartography, remote sensing, geographical education, geospatial analysis, techniques and modeling, landscape/regional and urban planning, economic geography, housing and the built environment, and quantitative geography. Volumes in this series may analyze past, present and/or future trends, as well as their determinants and consequences. Both solicited and unsolicited manuscripts are considered for publication in this series.

SpringerBriefs in Geography will be of interest to a wide range of individuals with interests in physical, environmental and human geography as well as for researchers from allied disciplines.

More information about this series at http://www.springer.com/series/10050

Rosalba D'Onofrio · Elio Trusiani

Urban Planning for Healthy European Cities

Springer

Rosalba D'Onofrio
School of Architecture and Design
University of Camerino
Ascoli Piceno
Italy

Elio Trusiani
School of Architecture and Design
University of Camerino
Ascoli Piceno
Italy

Additional material to this book can be downloaded from http://extras.springer.com.

ISSN 2211-4165 ISSN 2211-4173 (electronic)
SpringerBriefs in Geography
ISBN 978-3-319-71143-0 ISBN 978-3-319-71144-7 (eBook)
https://doi.org/10.1007/978-3-319-71144-7

Library of Congress Control Number: 2018931162

Printed on acid-free paper

This Springer imprint is published by the registered company Springer International Publishing AG part of Springer Nature
The registered company address is: Gewerbestrasse 11, 6330 Cham, Switzerland

Foreword: Resilience and Welfare Reform

This book, by Rosalba D'Onofrio and Elio Trusiani, addresses health and well-being in the city, which leads to a reflection on the reasoning of modern and contemporary urban planning, highlighting the distance that separates one from the other rather than their persistence.[1]

Questions pertaining to hygiene, as the authors remind us, are among the founding questions of urban planning, which originated with the "second urban revolution"[2] and fed different ideas of the modern city, moving progressively to a basis on ideas pertaining to welfare. Standards and zoning,[3] the two cornerstones of modern urban planning, which are at the centre of the disciplinary "review" today, form the technical response that European urban planners used to realize a healthy, just city over the course of about 80 years (starting with Ildefonso Cerda's general theory of urbanization published in 1867 up to the Athens Charter published by Le Corbusier in 1942). This was a universalist response, ethically founded and politically supported by movements and shared by socialist parties where it was taught starting from the needs of the mass of urbanized workers.

Reference to the "origins of modern urban planning"[4]—or perhaps better, to some of its "roots"[5]—provides a way to investigate the current explosion, which, faced with profound changes in the urban and environmental order, with evident effects on public health, affects the entire field of urban planning, redefining the themes, techniques, procedures, and tools.

---

[1] See Clementi, A. (2016). *Forme imminenti. Città e innovazione urbana*, LIStLab, Rovereto on the relationship between the modern and contemporary eras.

[2] The second and third urban revolutions in the sense given by F. Ascher, *I nuovi principi dell'urbanistica*, M. Russo (Ed.), Tullio Pironti editore, Naples 2005.

[3] The zoning reform, as it is known, is one of the doctrinal points in the Athens Charter. See Di Biagi, P. (Ed.) (1998). *La Carta d'Atene. Manifesto e frammento dell'urbanistica moderna*, Officina, Rome.

[4] Benevolo, L. (1991). *Le origini dell'urbanistica moderna*, Laterza, Rome-Bari.

[5] Secchi, B. (2007). *Prima lezione di urbanistica*, Laterza, Rome-Bari.

The original research presented in the central part of this book consists of a broad reconstruction of the operational framework, which follows European recommendations, the achievements of the Healthy Cities movement, and some significant door-opening experiences in European cities. As a whole, these experiences have given shape to a theme that, for the last decade, has merited careful critical attention. In particular, according to the authors, from the "laboratory" of the 1400 cities composing the Healthy Cities Network, "a new 'idea' of city, a new means of organizing functions in space, composing the urban form, organizing the city's relationship with the environment and the landscape—in effect, a new model of urban planning" would make inroads. In effect, the conditions required for cities to adhere to the Health 2020 strategy directly affect urban planning and design, identifying choices regarding land use, social services, and transport as some important cornerstones. Other important lines of work include policies of adapting to climate change and community resilience. In other words, by focusing on health and well-being threatened by ageing, chronic disease, and diseases transmitted by infection and urbanization,[6] cracks can be seen in the fundamental achievements of modern urban planning as well as in the limits of the urban-planning field. It is not by chance that the internal urban-planning debate addresses and also affects the way of dealing with historical questions (land use, social services, transport) and more recent questions (climate change, resilience).[7]

There are many implications situated on many different planes. I refer to only some of them as examples of actions ranging from housing ergonomics to supranational policies for different climate regions.

As a characteristic datum, new demographic conditions show an ageing population, with the trend in Italy that by 2025 will see positive balances in only 23 provinces, almost exclusively in the north.[8] Such a consistent, diffuse presence of people with various forms of disabilities related to advanced age requires a new set of city facilities, organized forms of service, and widespread accessibility, but also a massive operation to renovate buildings, especially considering solitude due to death or the distance from family. For example, installing lifts and removing architectural barriers within buildings is fundamental (which otherwise risks "reclusion", with its social and economic costs). This type of intervention, added to those for energy and static renovations, is one of the main reasons for regenerating existing buildings. On the other hand, the current profound reorganization of

[6] These are the "new challenges for health systems" delineated in the Preface to the book by D'Onofrio and Trusiani (2017) *Città, salute e benessere. Nuovi percorsi per l'urbanistica*, by Andrea Lenzi, President of the National Committee for Biosafety, Biotechnology, and Life Sciences under the Presidency of the Council of Ministers and the Health City Institute.

[7] See "Progetto Paese" (Country Project) presented by the Italian National Institute of Urban Planning at its 29th Conference (Cagliari 28–30 April 2016); the acts of the 19th Italian Society of Urban Planners Conference (Catania 16–18 June 2016; www.planum.net); La Biennale public space program (Rome, 25–27 May 2017).

[8] Data from CRESME (Italian Centre for Economic and Social Research in the Building Market), 2017.

social/health services related to home assistance is already proposing new forms of living based on different degrees of sharing. These are found above all in the area of social housing, but are beginning to come into view of the most up-to-date real estate investors. The impacts on transport, while obvious, still seem to be suffocating. The means of managing local public transport have still not been sensitized in this sense, while the relevance of continuous, safe (not slippery, for example) pedestrian paths is subordinate to cycling paths.

Among chronic diseases, cardiovascular disease in particular is associated with incorrect lifestyles, notably sedentariness. The short circuit with urban operation is clear, almost immediate: preventive therapies (movement outdoors in well-oxygenated areas) create a spatial organization that makes different modes of life possible and easy. Continuous, branching biking paths that intersect daily habits (so they are no longer and not only for leisure), comfortable, attractive, or dedicated walking paths (fitness courses), green spaces, and equipment for various recreational activities are already present in guidelines for the design of public spaces. In the health perspective, however, they find further reason for focusing designers' attention on more specific, refined solutions and a pervasiveness that borders on hygienic obsession. Even horticulture, a characteristic component of peri-urban agriculture, where the conditions have been less dramatically changed by the crisis and are less socially polarized, is confirmed in relation to the need to affirm new lifestyles and new value systems (attention for the environment, food quality, and a sharing economy).[9]

Diseases tied to urbanization refer not only to dust pollution mainly due to vehicular traffic (damage to the respiratory system and allergies in general), but also to noise pollution and smells, with the systemic consequences of the stress they can cause. American studies on the relationship between urban sprawl and health show a correlation between high street use and both the reduction of physical activity and the incidence of mental illness, including even a reduction in social capital. The studies highlight a specific, serious penalization of vulnerable populations (due to age, disability, skin colour, and income).[10] Conversely, D'Onofrio and Trusiani refer to the positive relationship between dense cities/public transport services and green areas with health.

Considering that the production of carbon dioxide is among the main causes of the global increase in temperature and considering that about 75% of CO_2 is produced in cities (distributed among housing, transport, and economic activities in roughly equal parts), we can include the impacts on health due to climate change among the consequences of urbanization. We can also add that environmental questions have entered the political agendas and subject to public attention precisely when they began to have evident negative effects on health. In the

[9] See, for example, Cinquepalmi, M., Petrei, F. (Eds.) (2015). *Ortipertutti. Nuovi orti a Bologna/New Gardens in Bologna*, Urban Center Bologna, Bologna.

[10] Frumkin, H., Frank, L., Jackson, R. (2004). *Urban Sprawl and Public Health: Designing Planning and Building for Healthy Communities*, Island Press, Washington (DC).

Mediterranean region,[11] specific threats to health arise from heat waves, the absolute lack and/or low quality of water, and the possibility of serious accidents due to extreme atmospheric events. Climate plans, which are now indistinguishable from plans for sustainable energy,[12] with their fusion of strategies and actions and their ability to interface with multiple urban plans, projects, and policies, constitute the technical product at the most interesting and precise moment to consider and contextualize questions pertaining to health and well-being.

It is precisely this scope of actions that I have summarily referred to and which explains the recommendation by the World Health Organization (WHO) to consider "Health in all Policies" that, in my opinion, advises against the introduction of a new tool aimed at assessing the impact of health (HIA), which would be added to the strategic environmental assessment (SEA) and the Italian environmental and territorial sustainability assessment (*Valutazione di sostenibilità ambientale e territoriale*, ValSAT). This relates not only to intolerance for the multiplication of specialized tools that are slowly making the content of policies opaque, but also their powerlessness with respect to the all-encompassing anxiety that inspires them or the willingness to control the *whole* that eludes us in sophisticated algorithms based on the identification and availability of powerful databases. One of the reasons for dissatisfaction with respect to the ValSAT, practiced for many years in support of urban plans, lies precisely in its specialized fragmentation. One loses sight of its relation to the problem since complexity is reduced by selecting some numerical indicators and the assessment of quality becomes a procedure, such that from denoting levels it risks being transformed into a flag to hold onto (as happened with standards). Paradoxically, tools designed to be integrated wind up fragmenting the framework of skills and lose the sense of the operations.

Without a doubt, there is a strong trend towards a new functionalist reductionism.[13] In this respect, the authors' open-ended conclusions, the reference to interpretation, culture, and skills to deal with extremely diverse contexts and circumstances are appropriate. Summary and determinism, in fact, are always risks lurking in disciplines with a low rate of specialization.

The importance and vastness of the theme posed by the book lead to questions about the universal character (or not) of the proposals, asking how it is possible to create cities that are healthy for all. While recognizing a general flattening in public discourse that supports a judgement of neo-hygienism, there are rivulets of reflection in its folds on the discrepancies and different impacts that different diseases have on populations. Even ageing is not "democratic", and not only for

[11] One of the seven climate regions in Europe according to the European Environment Agency report *Urban adaptation to climate change in Europe 2016. Transforming cities in a changing climate*, EEA Report, no. 12, 2016.

[12] As of 2017, the energy initiative promoted with the Covenant of Mayors is now integrated with the initiative on climate change in the Climate Change Adaptation through the Covenant of Mayors for Climate and Energy.

[13] Widely discussed by Cristina Bianchetti in her latest book, *Spazi che contano. Il progetto urbanistico in epoca neo-liberale*, Donzelli, Rome 2017.

economic reasons, given the importance assumed by social and cultural capital in a "hypertext society".[14] Cleaning up the air, land, and subsoil is now recognized as a condition for survival (of the human species through other living species) and, in this sense, is an objective of universal worth. However, levels of risk and threats are usually polarized, as already shown by the research on urban sprawl mentioned above. This awareness is decisive for the choice of priorities and the refinement of policies, to identify a thread in the extraordinarily intricate bundle of problems to be faced. There is a city of the rich and a city of the poor,[15] and there is a responsibility that is distributed among politicians, administrators, and those in charge of constructing the "urban agenda". The initial step is to not confuse the levels and then to not oppose the hygienic drift with a welfare drift. Instead, intersections among the different objectives should be identified to work on in depth, along with the possible confluences (with synergic effects) and dilemmas when objectives and actions aimed at realizing healthy cities run into objectives and actions aimed at building just cities. It seems to me that the effective encounter with the different contexts is situated precisely at this intersection and community resilience can be expressed. For this reason, it is not reasonable to quash social questions with environmental questions or maintain that policies for environmental resilience respond simply to the need for a new welfare. Environmental and social instances meet but do not identify each other, and devices to redistribute spatial richness do not coincide with those for mitigation and adaptation to climate changes if none other than for different time horizons and, in many cases, due to the competition regarding the destination of resources.

Milan, Italy

Patrizia Gabellini
Politecnico di Milano

[14]As defined by Ascher, in contrast to the industrial society, op. cit.

[15]Secchi, B. (2013). *La città dei ricchi e la città dei poveri*, Laterza, Rome-Bari.

Preface

In 2012, the Lancet Commission conducted a study into potential innovative associations between issues of health, social (in) equality, and economic development in city planning. This study recognizes the so-called urban advantage for human health and focuses on limitations of the linear and cyclical approaches to urban planning in dealing with the issues of health and quality of life of city inhabitants. In doing so, the Commission expressed the belief that urban planning is the most appropriate tool to move from the rhetoric of many policies aimed at promoting health and safety in the city to practical actions. The study requires planning to focus on experiments and projects while involving local communities and planning at various levels. More recently, the UCL-Lancet Commission 2015 report "Health and Climate Change" says climate change could be the greatest global health opportunity of the twenty-first century and it encourages the transition of cities to promote and support lifestyles that are healthy for both individuals and the planet.

This book uses the above as a starting point and aims to investigate different aspects of European Healthy Cities, examining various best practices. Capitalizing on ongoing trials, the book identifies the policies that underlie plans and projects that have caused positive changes in local communities in terms of the quality of life, health, and well-being of inhabitants. From these best practices, the book deduces some themes, strategies, and general criteria for planning healthy European cities.

The book is organized into three parts.

PART I—The City for Better Living

With reference to the international literature, the first part of the book addresses the different aspects of healthy cities, evaluating synergies with other interesting issues concerning contemporary cities. It describes the successes and failures of the European Healthy Cities Network. Finally, it lists the main inspiration for new urban governance to promote the well-being and health of European cities. This first part includes contributions from two cities: Belfast and Bologna, experts on health, city well-being, and the governance of urban phenomena.

PART II—Healthy Urban Planning in Europe

The second part investigates the role of urban planning in promoting concrete actions to improve the quality of life, health, and well-being in the city. This was done through a selection of some practices in different European cities, with the aim of identifying and investigating relationships between: (a) health promotion and urban sustainability; (b) possible conflicts and synergies between different levels of urban policies and between different urban actors and local communities; and (c) technical and operational tools that cities have implemented to ensure public health. The cases investigated include cities such as: Belfast, Bologna, Bristol, Copenhagen, Poznań, Rennes, Rotterdam, Turin, and Turku.

PART III—Planning and Designing Healthy Cities and Communities

Based on European and international experiences, the third part defines strategies and criteria to reformulate and adapt urban plans and projects aimed at building health-friendly urban environments. First, it promotes the assumption of neighbourhoods as an ideal field of action to understand the challenges to health and well-being, intercept and stimulate the participation of local communities, and understand the design aspect of the planning choices that are increasingly tied to the quality of life, health, and well-being of the citizens. Second, the exploratory role of the project is considered in order to reposition and reorganize urban spaces with respect to the potential impacts of the transformations and effects due to climate change on health and well-being of city inhabitants. Recourse to checklists, guidelines, and design orientations is established, which can be of assistance in stimulating discussion and negotiation among the different actors on the urban scene and in local communities. Finally, in this dimension, urban design takes on two new meanings among the most debated aspects in contemporary urban planning: densification and the temporary nature of city uses. In particular, the former appears as a sort of prerequisite for some recommended actions in terms of health, such as walking, socializing, sharing spaces. The latter serves as an occasion to approximate the quality design choices over time in an attempt to contribute to creating healthier and more equitable places and lifestyles.

Ascoli Piceno, Italy

Rosalba D'Onofrio
Elio Trusiani

Acknowledgements

For their useful collaboration and exchange of ideas, we would like to thank:

- the City of Bologna: Luca Rizzo Nervo (Assessor for Health, Welfare, Social and Solidarity Innovation) and Maria Cristina Zambon (Healthy and Healthy City Manager)
- Belfast Healthy Cities: Jonna Monaghan, Laura McDonald, and Joan Devlin
- the City of Udine: Furio Honsell (Mayor of the City of Udine); Stefania Pascut (WHO “Healthy City” Promotion Office); Raffaele Shaurli and members of the Urban-Planning Management organizational unit.
- Andrea Lenzi, President of the National Committee for Biosafety, Biotechnology, and Life Science under the Presidency of the Council of Ministers and the Health City Institute.
- Laurine Tollec, Ministère des Affaires sociales et de la Santé, France.
- Frédéric Auffray, Rennes Métropole.
- Szymon Błażek (Department of Real Estate Management City of Poznań)
- Wouter Vos—KuiperCompagnons
- Gaby Kurth—ISOCARP
- Joe Brickley—GIS Lead Officer at Bristol City Council Parks and Green Spaces
- Giuseppe Losco (Director of the School of Architecture and Design at the University of Camerino)
- Maurizio Maria Sani, former Director of the Emilia Romagna Region (Italy)

Particular thanks go to Prof. Patrizia Gabellini (Milan Polytechnic University), who wrote the preface to this book, our young colleagues, who wrote the city reports (Chiara Camaioni, Piera Pellegrino, Michela Tolli, and Flavio Stimilli), and finally Dr. Michelle Ouellette, our invaluable, patient, Italian-English translator.

Contents

Additional materials of this book are:

Appendix 1—Expert knowledge

Creating healthy and child friendly places in Belfast by Jonna Monaghan and Laura McDonald, Belfast Healthy Cities

Bologna_Healthy City: Social Sustainability, Urban Projects, and Innovations in Agreements by Luca Rizzo Nervo

Appendix n.2—City Experiences

Bristol's parks and green space strategy by Piera Pellegrino

Rennes-Restructuration de la halte ferroviaire de Pontchaillou by Michela Tolli

Healthy Poznan (HP)—Health Development Plan for the City of Poznan by Flavio Stimilli

The Hirvensalo District Master Plan: health impacts of three structural models in Hirvensalo—Turku by Chiara Camaioni

Part I
The City for Better Living

Chapter 1
For the "Human" Development of Cities in an Era of Climate Change

Abstract The connection between health, well-being, and the quality of living spaces is not accidental. The organization of the city and, in general, of social and environmental contexts, is capable of conditioning and modifying emerging needs, lifestyles, and individual expectations. Faced with scientific evidence for these relationships, it is necessary for urban planning to realize that there is no time left to hope that economic growth and demographic change, by themselves, will be able to generate conditions conducive to people's quality of life. This invitation is energetically shared by the WHO's Healthy Cities Movement. Through an interdisciplinary group that met between 2009 and 2011, the UCL–Lancet Commission developed a series of recommendations for policy makers to improve the urban environment and to open a discussion on the role that urban planning can play.

Keywords Healthy cities movement · Urban human scale · Well-being
Quality of living spaces · Climate change

A 2014 report on CNN coined the slogan "…Our health is not just a by-product of how we live. It's also about where we live". This was based on the recognition that leaders in cities like Copenhagen and Okinawa, as well as Vancouver, Melbourne, New York, etc., had recently implemented urban policies to provide their citizens with healthy food, access to parks, good public transport, disease control, and assistance for vulnerable segments of the population. In the same report, it was observed that the success of these initiatives was not only to be found in good policies, but also in citizens' dedication to promoting them with their everyday behaviour.

Some years before, in 2011, in the documentary film *The Human Scale*, the Danish architect and professor Jan Gehl had argued for the need to recover the "human scale" in building cities, hoping that they would be built around people rather than technologies.

These two different voices invite researchers, technicians, and politicians to reflect on how the design of the urban environment influences health, well-being, and the quality of life in cities, and on the need to increase knowledge of this relationship and encourage physical designs for the urban space that deal with these aspects.

R. D'Onofrio and E. Trusiani, *Urban Planning for Healthy European Cities*, SpringerBriefs in Geography, https://doi.org/10.1007/978-3-319-71144-7_1

The connection between health, well-being, and the quality of living spaces is not accidental. The organization of the city and, in general, of social and environmental contexts, is capable of conditioning and modifying emerging needs, lifestyles, and individual expectations. Until some decades ago, this opinion pertained to the "common feeling" of people; today it is supported by numerous studies and research. Scholars are particularly interested in the implications of the social context and characteristics of the urban space (Sampson 2003; Helliwell and Putnam 2004) because the research shows that their role is fundamental in determining—in both good and bad ways—the health and well-being of the community (Duhl and Sanchez 1999).

This was demonstrated in a 2013 study at the European Centre for Environment and Human Health at the University of Exeter, which explored the relationship between green areas and well-being. Based on a program that involved 10,000 participants over 18 years, it was verified how on average, individuals experience less mental discomfort and a higher level of well-being when they live in urban areas where there is significant green area. The program also highlighted how, while the effects on the individual level are important but not elevated, the potential overall benefit on the community level is substantial (White et al. 2013).

But this is not only about green areas. Another research project, this time from the University of Warwick, quantified the impact of scenic environments on health. According to the researchers, the aesthetics of the environment in which we live has quantifiable effects on our well-being, and harmonious architecture and design also produce a positive effect that is even more significant than the presence of green areas (Seresinhe et al. 2015).

Both of these studies highlight the need and opportunity to adopt adequate devices when designing urban spaces because their quality is related to our well-being and health. This is also the conviction of the World Health Organization (WHO) in reference to urban planning and its role of primary prevention, which contributes to good health (Duhl and Sanchez 1999). In connecting health to the urban dimension, health as an "individual good" becomes health as a "collective good", recalling the ethics and observance of rules of civil coexistence. Health becomes an objective for citizens, mayors, and local administrations to pursue and should be proposed as guaranteeing an equitable city, ensuring that community health is considered an investment and not a cost. The health-based city becomes a social and collective result, the result challenging globalization, social exclusion, and poverty.

The European Charter for the Safeguarding of Human Rights in the City, signed in 2000 by about 350 European cities (today numbering more than 400), identified the right to health, environment, and harmonious urban planning with some of the fundamental inspiring principles for European cities.[1] These principles were introduced in the document in a non-random sequence, almost to underline their close interrelation and consequentiality.

Faced with scientific evidence for these relationships, it is necessary for urban planning to realize that there is no time left to hope that economic growth and

[1] Art. XVII sets out the cities' commitment to promoting actions in the economic, cultural, social, and urban planning areas to promote health for all inhabitants, based on their active participation.

demographic change, by themselves, will be able to generate conditions conducive to people’s quality of life. On the contrary, there is time for openness to experimentation. The risk factors for health and well-being should become important variables in activities to design modern cities.

This invitation is energetically shared by the WHO’s Healthy Cities Movement. This movement was created in Toronto (Canada) in 1984 at the Beyond Health Care Conference with the objective of engaging local authorities in health development through a process of political commitment, institutional change, capacity-building, partnership-based planning and innovative projects. In more than thirty years, it has increased awareness that risks to health in urban environments are not being addressed appropriately (Kenzer 1999). Today, however, more than understanding how these risks can influence the health of city inhabitants, it aims to understand how well-planned and well-designed cities can produce benefits for health, as underlined in the WHO’s declaration of 2010 as the Year of Urban Health.

Through an interdisciplinary group that met between 2009 and 2011, the UCL–Lancet Commission developed a series of recommendations for policy makers to improve the urban environment and to open a discussion on the role that urban planning can play (Rydin et al. 2012).

These recommendations are based on the wide definition of health set out by the WHO in 1948: “Health is a state of complete physical, mental and social well-being and not merely the absence of disease or infirmity”. The recommendations clearly state that:

- public health should necessarily be the object of interdisciplinary work. There is a particular need for an alliance between urban planners and experts in the health sector;
- in planning and designing the urban environment, a key objective should regard the elimination of social inequalities and address access to health services between the different urban areas of a given city;
- the city should be modified to maintain the so-called “urban advantage for health”, identifying new points of reference for urban planning;
- political responsibility on the national and local scales are particularly important for understanding the complexity of the theme of health and the overlapping of roles and skills that influence urban policies, as well as the effects of these policies on the health of city inhabitants;
- the effectiveness of actions in matters of health is pursued through experiments and designs on the local scale. These activities necessarily involve local communities and interest holders.

In particular, the Commission pointed to a gap between aspirations and outcomes in terms of urban and environmental health with a warning. This is because the presumed achievements of the city (urban areas have greater resources, better infrastructure, and a wider availability of services than rural areas) are difficult to preserve and implement over time. In addition, the first WHO–UN Habitat report of 2010, “Hidden cities: Unmasking and overcoming health inequities in urban settings”, highlighted that even where the prosperity of cities is increasing, there is always a "hidden" side. This relates to poverty in the most rundown neighbourhoods, even in the richest cities in the world. Continuing down this road, there is a risk of seriously blocking the objectives of development established by the new Sustainable Development Goals (SDGs) to stop poverty, protect the planet, and ensure prosperity for all (UN 2016).

Although generalizations cannot be made, the WHO's 2016 Global Report on Urban Health suggests that tested solutions exist to address the challenges of health and well-being. Progress in this direction has not only regarded the efficiency of health services, but also the capacity to shape urban environments (WHO 2016). If it is in fact true that "…Not every city can do an 'extreme makeover' for health", it is also true that "...every city can take steps in the direction of healthier planning".

According to the WHO's report, working in this direction means several things: making daily places easily accessible; interpreting the theme of urban compactness and density in an innovative way, reasoning about the composition of spaces and functional *mixité*; making cities age-friendly; and rethinking cities so that they become more resilient to the impacts of natural phenomena and climate change (including floods, earthquakes, urban heat islands, droughts, fires, etc.).

These impacts can really test both infrastructures and human health, as stated in the IPCC's Fourth Assessment Report. There are three main mechanisms by which climate change may affect human health: direct exposure to extreme climate events; indirect effects from changes to the determining factors of human health; and effects of climate events on social welfare by disrupting social and economic systems (Parry et al. 2007).

Combining mitigation, adaptation, and health strategies constitutes the challenge for a transition towards a more sustainable, healthy society. In this challenge, cities can offer "…unique opportunities to marshal resources and wealth to build resilience and health-protective policies and programs" (Barata et al. 2011). However, it is necessary to be aware that health-care adaptation measures will be different from city to city because the social, economic, cultural, and political realities are different. However, the basic objectives should be shared by all for the safety of cities and our own safety.

References

Barata, M., Ligeti, E., De Simone, G., Dickinson, T., Jack, D., Penney, J., et al. (2011). Climate change and human health in cities. In C. Rosenzweig, W. D. Solecki, S. A. Hammer, & S. Mehrotra (Eds.), *Climate change and cities: First assessment report of the urban climate change research network* (pp. 179–213). Cambridge: Cambridge University Press.

Duhl, L. J., & Sanchez, A. K. (1999). Healthy cities and the city planning process: A background document on links between health and urban planning. Copenhagen, WHO Regional Office for Europe. http://www.euro.who.int/document/e67843.pdf. Accessed May 31, 2017.

Helliwell, J. F., & Putnam, R. D. (2004). The social context of well-being. *Philosophical Transactions of the Royal Society of London. Series B, Biological Sciences, 359*(1449), 1435–1446. https://doi.org/10.1098/rstb.2004.1522.

Kenzer, M. (1999). Healthy cities: A guide to the literature. *Environment and Urbanization, 11*(1), 201–220. http://journals.sagepub.com/doi/abs/10.1177/095624789901100103. Accessed May 31, 2017.

Parry, M. L., Canziani, O. F., Palutikof, J. P., van der Linden, P. J., & Hanson, C. E. (2007). *Contribution of working group II to the fourth assessment report of the Intergovernmental Panel on Climate Change*. Cambridge, UK and New York, NY, USA: Cambridge University Press.

Rydin, Y., Bleahu, A., Davies, M., Dávila, J. D., Friel, S., De Grandis, G., et al. (2012). Shaping cities for health: Complexity and the planning of urban environments in the 21st century. *The Lancet, 379*(9831), 2079–2108.

Sampson, R. J. (2003). The neighorhood context of well-being. *Perspectives in Biology and Medicine, 46*(3 Suppl), S53–S64 (Johns Hopkins University Press).

Seresinhe, C. I., Preis, T., & Susannah Moat, H. (2015). Quantifying the impact of scenic environments on health. *Scientific Reports, 5,* 16899. https://doi.org/10.1038/srep16899.

UN. (2016). Sustainable development goals. 17 Goals to transform our world. http://www.un.org/sustainabledevelopment/sustainable-development-goals/. Accessed May 31, 2017.

White, M. P., Alcock, I., Wheeler, B. W., & Depledge, M. H. (2013). Would you be happier living in a greener urban area? A fixed-effects analysis of panel data. *Psychological Science, 24*(6), 920–928. https://doi.org/10.1177/0956797612464659.

WHO. (2016). *Global report on urban health: Equitable, healthier cities for sustainable development*. http://www.who.int/kobe_centre/measuring/urban-global-report/en/. Accessed May 31, 2017.

WHO, UN HABITAT. (2010). Hidden cities: Unmasking and overcoming health inequities in urban settings. http://www.who.int/kobe_centre/publications/hidden_cities2010/en/. Accessed May 31, 2017.

Chapter 2
Goals, Opportunities and Limits to the European Healthy Cities Network

Abstract The history of European cities reflects the close, complex ties that unite urban planning and human health. An effective remedy against epidemics in the nineteenth and early twentieth centuries, city planning has, paradoxically, contributed to the appearance of many problems related to the health and well-being of people in the modern era. In September 2012, the European Member States of the WHO adopted "Health 2020", a strategic policy framework for the twenty-first century. Health 2020 explicitly recognizes the influence of the urban environment on health and the role of healthy cities and national networks in carrying forward the objectives and themes of this European strategy. Health 2020 also recognizes the emblematic role of the leadership of local governments in the development of health. Urban planning should address this activism in European cities and the need to overcome what can be defined as the risk of "projectism", the risk of a short-term vision relying on isolated interventions rather than long-term programs or policies that can profoundly modify the organization of contemporary urban models in favor of the health and well-being of city inhabitants.

Keywords Urban health · Health 2020 · Local governance · Long-term urban policies · Change of contemporary urban models

2.1 For a New Alliance Between Health and City

The history of European cities reflects the close, complex ties that unite urban planning with human health. An effective remedy against epidemics in the nineteenth and early twentieth centuries, city planning has, paradoxically, contributed to the appearance of many problems related to the health and well-being of people in the modern era.

The separation of urban functions theorized in 1933 in the Athens Charter into "dwelling, work, recreation (use of leisure time), transportation" has led to zoning, urban development based on automobile transport, and the spread of new individual behaviours. In turn, this has led to problems for the environment and health, such as

R. D'Onofrio and E. Trusiani, *Urban Planning for Healthy European Cities*, SpringerBriefs in Geography, https://doi.org/10.1007/978-3-319-71144-7_2

atmospheric pollution, an increase in noise, reduced physical activity, illnesses due to changes in eating habits, the loss of social ties, isolation, marginalization, etc.

Faced with the unsustainable nature of these changes, it is time that cities turn the page, making a great effort to design strong, healthy, and vibrant places and requiring that the discipline of urban-planning has a renewed focus in favouring this new phase in urban history.

The links between health and urban planning are not new. They have evolved over time with reference to three main events: the period of hygiene applied to urban planning in the nineteenth century, functional urban planning up to the 1970–80s, and finally the current period of "sustainable" urban planning (Roué-Le Gall et al. 2014). Each of these periods corresponds to choices in European urban planning that were made in response to demands related to health.

As a background article on the sustainable city by the Danish Architecture Centre (DAC 2014) begins, "Health is [currently] one of the fundamental prerequisites for a sustainable lifestyle in the cities". It is also "…a precondition for sustainable development" (Hancock 1993; Kickbusch 2010).

These two affirmations shed new light on the relationship between sustainable development and human health, which is too often trampled by a rather restrictive environmental approach that focuses primarily on conserving environmental resources, assessment, and controlling the risks of urban transformations on health. If it is in fact true that promoting health consists in protecting the environment and minimizing the existence of risks contained therein, the risks cannot be ascribed exclusively to assessing environmental pressures. Nor, on the other hand—still in reference to environmental pressures—is it plausible that a project for urban development is, by itself, incompatible with the need to preserve the environment and health.

On the contrary, it is precisely the need to find compatible solutions among city models, quality of life, and the health and well-being of citizens that implies a search for an integrated approach, involving different skills and actors, setting aside the approach of sectoral boundaries/silos (de Leeuw and Green 2017). This traditional approach addresses the theme of health in a simplistic way according to a linear cause-and-effect procedure that tends to isolate individual variables in a reductionist manner (de Leeuw 2011). Working with the complex theme of health also requires an adaptive, flexible approach capable of creating new connections and relationships over time (Kickbusch 1999). This was implied by the WHO in the Ottawa Charter of 1986, which established the Healthy Cities movement in Europe. It defines the promotion of health as a "…process of enabling people to increase control over, and to improve, their health. To reach a state of complete physical, mental and social well-being, an individual or group must be able to identify and to realize aspirations, to satisfy needs, and to change or cope with the environment. Health is, therefore, seen as a resource for everyday life, not the objective of living".

This concept of health as a process involving different subject and aspects rather than a product is at the basis of the "requirements of engagement" that the WHO European Healthy Cities Network (WHO/EHCN) uses when selecting the cities for the network. Each city should also accept and dedicate itself to realizing these requirements over time (Tsouros 2015). The requirements of engagement, borrowed from a series of key policy documents, have slowly been reinforced in the direction of promoting health for all, equity in health, sustainable development, and good governance.

From its establishment, the EHCN has evolved through a series of working programs called phases. These have served as a platform for inspiration, learning, and the accumulation of practical experiences on how to improve health and well-being in cities. They have been useful for measuring the progress of urban policies over time and to establish priorities. In September 2012, the European Member States of the WHO adopted "Health 2020", a strategic policy framework for the twenty-first century. This framework is based on the values and principles expressed in "Health for All" and other key health policies developed in the last decade. Health 2020 explicitly recognizes the influence of the urban environment on health and the role of healthy cities and national networks in carrying forward the objectives and themes of this European strategy. Health 2020 recognizes the emblematic role of leadership of local governments in the development of health (WHO Regional Office for Europe 2013a).

In the current phase of the program—Phase VI (2014–2018)—the general objectives and themes of the Health 2020 strategy are being applied to the local context based on four priority actions (WHO Regional Office for Europe 2013b):

- investing in health throughout the course of life (life-course approach) and empowering people;
- tackling major public health challenges;
- strengthening people-centred health systems and public health capacity;
- creating resilient communities and supportive environments.

With regard to the last priority action, the Official Call for Expression of Interest has identified some main themes for this new phase:

- *Community resilience*. Construction and promotion of resilient urban actions to improve health on the individual and collective levels. Cities should create healthy environments and favour the empowerment of people so they can make healthier choices.
- *Healthy environments*. The constant improvement of living and working conditions are a fundamental component of local development. Cities should promote actions to improve places where people live, work, and play, such as housing, schools, workplaces, hospital structures, and care homes.
- *Urban planning and design according to health criteria*. Living conditions in cities influence health through the built environment, the social environment, and access to services. Improvements can be made with urban planning that favours the practice of physical activity and the development of sustainable mobility.
- *Transport that favours healthy lifestyles*. A good public transport service, together with biking and walking paths, can notably reduce pollution, noise, energy consumption, and traffic congestion, improve road safety, protect the landscape and urban cohesion, provide greater opportunities to practice physical activity, and

improve access to social, educational, recreational, and professional services in the city.

- *Climate change*: involvement in demonstrating the existence of a strong link between sustainable development and health in support of policies to mitigate and adapt to climate change.
- *Interventions on housing and urban regeneration*. Great health benefits can be obtained by intervening in the building sector through a series of measures that favour health, such as active and passive ventilation for cooling, interventions to reduce humidity, more efficient and less polluting heating, better health and hygiene conditions in buildings, etc. Urban regeneration programs can also provide greater social, economic, and environmental opportunities and can contribute to reducing inequality in cities.

All of these themes are central to urban planning, which is called to respond to the new demands of a community that is finally more attentive to the themes of health and well-being.

2.2 "...Thirty Years After the Birth of the Movement"

The Healthy Cities movement has been active in Europe for thirty years now. Nearly 100 cities are members of the WHO European Healthy Cities Network, and 30 national Healthy Cities Networks across the WHO European Region count more than 1400 cities and towns as members. Over time, cities in the network have become true laboratories where innovative approaches for health and sustainable development have been developed and expanded (Tsouros 2017). After thirty years, we should be able to determine the efficacy of programs promoted by the movement to improve the city as a living environment.

Twenty years after the creation of the network, in his 2009 book *Toward the Healthy City–People, Places, and the Politics of Urban Planning*, Jason Corburn defined some of the network's limits with the support of expert opinion. The first sign, in reference to a claim made by Takano in 2003, regarded the limited attention for the combination of "policy processes, science norms, and organizational network building" (Takano 2003). In agreement with de Leeuw and Skovgaard (2005), Corburn also underlined the lack of success of actions promoted by the network regarding the effective ability to influence urban plans and the way of making choices related to develop, design, and manage cities.

The debate around the efficacy of actions proposed by the movement was reignited some years later, in 2015, by the magazine *Health Promotion International*. The magazine dedicated a supplement of volume 30 to evaluating the WHO/EHCN's Phase V (2009–2013) and, with the contribution of different authors, synthetically retraced the different phases of implementing the program, the innovations made, and the difficulties encountered.

With the aid of this publication and drawing from the rich bibliography available, the salient characteristics of the different phases are briefly described below, focusing attention on the evolution of the health/urban planning relationship.

While the Healthy Cities program began to involve matters of urban planning from the end of the 1980s, it was Phase III (1998–2002) that placed great importance on the need to integrate the objectives of health within urban planning. Its activation in 1998 began with a questionnaire given to the 38 cities participating in Phase II (1993–1997). This questionnaire was aimed at administrators in the departments of urban planning in the different cities. The results showed that in only 25% of cases was real cooperation between health and urban-planning policies registered. Nearly a third of interviewees said that the urban policies developed by the cities were incompatible with health, although cities had registered different problems falling under the strict discipline of urban planning, such as excessive levels of motorized traffic, social segregation, and a lack of attention to citizens' daily needs (Barton and Grant 2011). The survey also showed that theory and practice had become significantly detached. The need to promote greater integration among the principles of the movement and experimentation in the field was the central theme of debate at a seminar held in Copenhagen in 1999, whose results were compiled in a WHO report and then in the much-cited book by Barton and Tsourou (2000). A working group was formed in 1999 to resolve this gap, adopting twelve key objectives that were to inspire Healthy Urban Planning (HUP) and were identified in close relation to those analogous to sustainable development and Agenda 21 (UN 1993). The objectives of HUP were to:

- promote healthy lifestyles (especially regular exercise);
- facilitate social cohesion and supporting social networks;
- promote access to good-quality housing;—promote access to employment opportunities;
- promote accessibility to quality services (education, culture, leisure, retail sales, and health assistance);
- promote the use of local food products and businesses for healthy food;
- promote safety and a sense of security;
- promote equity and the development of social capital;
- promote an attractive environment with acceptable levels of noise and good air quality;
- guarantee good water quality and good hygiene/sanitary services;
- promote the conservation and quality of the terrain and mineral resources;
- reduce emissions that threaten the stability of the climate.

Based on experiments made in six cities, the different possible levels of integration between urban planning and health were also classified (Taylor 2010). The first level regards the recognition of elements that are essential for settlements: the existence of shelter, access to food and water, air quality, and wastewater treatment. These conditions are taken for granted in Western Europe. The second level goes beyond environmental health. It is recognized that many aspects of settlement systems and their design influence the health and well-being of citizens, such as parks, gardens, biking paths, reduced dependence on cars, and the low emission of pollutants. This level addresses questions further down the planning process and therefore relate to partial integration. The third level, which is rarer, regards active collaboration among the different sectors and administrative areas to design a healthy city together with planners.

Phase IV (2003–2007) of the Healthy Cities program views HUP as one of the main themes that all member cities in the network should have adequately developed based on the twelve objectives in the preceding phase. They should be appropriately reviewed and integrated over time with biodiversity, food, energy, and waste.

Evaluation of this phase was made based on a General Evaluation Questionnaire (GEQ) and Annual Reporting Templates (ARTs). Fifty-one cities responded to the questionnaire out of the 77 adhering to the network. The questions aimed to investigate the degree to which city planning was in line with the strategic priorities of HUP and which of the strategic priorities identified was deemed the most important. The results of this assessment (Barton and Grant 2011) revealed that:

- Nearly 65% of cities declared themselves to be actively involved in promoting the HUP; another 20% were aware of policies in these fields and shared them, while being not very involved. A small minority (15%) confirmed that they were still not working in that direction.
- When addressing themes such as inequality and equity, in only a few cases was it possible to find answers in urban planning policies. For only 25% of those interviewed was this interrelationship clear; 43% never mentioned these aspects; and 35%, while being aware of this interrelationship, could not find it in current policies.
- In reference to the quality of activities[1] aimed at promoting urban planning favourable to health, the level reached by cities has increased over time. The number of cities placed in the high level in 2005 (11) grew to 26 cities in 2008.

These assessments show rather clearly how in many cities, the integration between health and urban planning policies requires fundamental changes in the organizational structure and management of responsibility. It points to the need to promote interaction and the exchange of knowledge between professionals in public health and urban planners (Pilkington et al. 2008).

The methodological path used to evaluate the first phases of implementing the program was deemed to be not very relevant by some scholars. de Leeuw and Green (2017) wrote explicitly about "inconvenient assessments (drawbacks)" for a series of reasons, among which are: a limited geographical scope, neither evidence- nor theory-based, etc.

---

[1]The indicators chosen for this assessment are whether the city:

- Addressed the twelve HUP objectives;
- Demonstrated integration with Healthy Impact Assessment and Healthy Ageing;
- Displayed a range of activity at different spatial scales;
- Evidenced both an integrated strategic approach and implementation at the local level;
- Involved a good range of relevant planning agencies and community stakeholders.

Table 2.1 Summary core designation criteria for Phase V

1. *Health and health equity in all policies*
2. *Caring and supportive environments* Subtopics: Better outcomes for all children; age-friendly cities; migrants and social inclusion; active citizenship; health and social services; health literacy
3. *Healthy living* Subtopics: Preventing non-communicable diseases; local health systems; tobacco-free cities; alcohol and drugs; active living; healthy food and diet; violence and injuries; healthy settings; well-being and happiness
4. *Healthy urban environment and design* Subtopics: Healthy urban planning; housing and regeneration; healthy transport; climate change and public health emergencies; safety and security; exposure to noise and pollution; healthy urban design; creativity and liveability

The need for an evaluative approach rather than "realist synthesis" [2] guided Phase V in assessing the project. This phase was centred on the following "core designation criteria": (1) Health and health equity in all policies; (2) Caring and supportive environments; (3) Healthy living; and (4) Healthy urban environment and design (Table 2.1). The novelty of the approach followed in this phase lies in the awareness that to demonstrate the effectiveness of policies regarding well-being, health, and social equity, an assessment approach was negotiated directly with the cities, in the knowledge that some policies require long-term investments and a long time to demonstrate results (WHO Regional Office for Europe 2014).

The comparison made on this basis showed (de Leeuw et al. 2015) that in Phase V of the movement, independent of the geographical location of the network cities:

- there was a move from small-scale, time-limited projects dedicated mainly to improving lifestyles for health to broader policies and programs involving good governance of health that worked and rested on questions of urban sustainability, urban planning, and equity;
- cities began to connect the dots between different interventions, questioning the results obtained by the policies implemented, with reference to social determinants of health[3], governance, and equity. This occurred even with the involvement of

[2]This consists in putting together a series of data from the 99 cities in the European network and the 31 national networks. These data regarded: annual data that each member city in the network should provide based on a unitary format; an online questionnaire, the organization of information according to three types of case study (thematic, on core themes of city status; strategic, on core attributes of healthy city activity; and proudest achievements); the assessment of indicators from Eurostat and national databases; and document analysis.

[3]Determinants of health are factors that influence the state of health of an individual, a community, or a population. They can be grouped into different categories: personal behaviours and lifestyle; social factors that can be an advantage or disadvantage, working and living conditions; access to health services; general socioeconomic, cultural, and environmental conditions; genetic factors. The Dahlgren-Whitehead model, which is widely used in Europe, reflects the European culture of the welfare state based on the "right to health" and adopts the "multi-sectoral" vision of protecting health contained in the Declaration of Alma Ata. This model contains a series of concentric layers

and consultation with stakeholders to develop visions and strategies that in turn, even without explicit recognition, are capable of building the objective of "Health in All Policies", which is the ultimate objective of the WHO.
- local governments are still suffering the effects of the global financial crisis and often operate under severe austerity. However, this does not prevent pursuing broad inter-sector strategies for the health of cities. A fundamental reason for this long-lasting vision is a strong recognition of the role that local communities play in influencing the development of policies and projects for health.
- the network is very active in reducing inequality in terms of health by promoting policies and programs based on actions proposed to achieve practical results and growing trust in Health Impact Assessments (HIA).
- the cities involved believe firmly that following the Healthy Cities Network approach makes a difference and that they are on the right path to contribute to a better quality of life for their inhabitants, but that other efforts still need to be made.

Based on the experiences of Phase V, Phase VI places greater attention on understanding the characteristics of the different urban contexts and their different impacts on the policies implemented in favour of health and well-being (Kickbusch and Gleicher 2012). The phase is also configured as an "adaptable and practical framework for delivering Health 2020 at the local level" (Tsouros 2017).

In fact, the two strategic objectives of Health 2020 (improving health for all and reducing health inequalities; improving leadership and participatory governance for health) provide the general framework within which the Phase VI is organized.

Starting from the premise that each city is a *unicum*, there can be many different approaches to the theme of health. In contrast, the objectives organized around a series of core themes[4] should be identical.

In choosing the core themes, there is a renewed interest and attention for aspects of health and well-being such as: healthy ageing; social inclusion; health literacy; physical activity; childhood obesity; dealing with stress, depression, and alcohol abuse, etc.; community resilience in dealing with both social and physical environmental challenges; healthy urban planning and design; and climate change, especially in terms of disaster preparedness and response.

---

corresponding to different levels of influence. At the centre there is the individual, with his or her biological characteristics: sex, age, genetic history, i.e., the determinants of health that cannot be modified. The modifiable determinants, those that can be corrected and changed, are situated in layers from the inside to the outside: individual lifestyles, social and community networks, living and working environment, the political, social, economic, and cultural context.

[4]Theme 1: Investing in health throughout the course of life (life-course approach) and Theme 2: Addressing the great challenges in public health in reference to both transmittable and non-transmittable diseases, favouring the empowerment of people; Theme 3: Reinforcing health systems centred on the individual and their capacities, short reaction time in emergencies, and surveillance in terms of public health; Theme 4: Creation of resilient communities and environments favourable to health.

The method used to assess the success of Phase VI must consider the means used by Phase V and should probably be reinforced by better clarifying the community's role throughout and beyond the process, from the initial expert framing of the problem to the final judgement regarding what works.

2.3 The Need to "Look Beyond"

Despite the effort to assess the different phases of the program's implementation and the objective difficulties in doing so, the key question is whether at the end there is reassuring proof that the movement's activities over time have helped to improve the health of city inhabitants and to address social inequalities.

This is the question asked by Helen Wilding and other authors in the conclusion of their contribution to the book *Healthy Cities. The Theory, Policy, and Practice of Value-Based Urban Planning* (de Leeuw and Simos 2017). The response given by this group of scholars is that "...answering this question may not be possible" for a series of reasons. These range from the methodological difficulty inherent in the evaluation, which is found when comparing very different local and national contexts, to the consideration that day-to-day practices in the cities can be very different from what was theorized. While it is certainly not possible to boast success in terms of "profits" in health and that social inequalities continue to be an emergency throughout Europe, there is growing interest in health and well-being in the Urban Agendas of many cities and incessant experimentation in the field to create better urban conditions.

Urban planning should address this activism in European cities and the need to overcome what can be defined as the risk of "projectism" (Goumans and Springett 1997), i.e., the risk of a short-term vision relying on isolated interventions rather than long-term programs or policies that can profoundly modify the organization of contemporary urban models in favour of the health and well-being of city inhabitants.

If we had to use two words to describe the fields in which it is necessary to work to promote a strategic long-term vision, these would undoubtedly be *governance* and *management*.

Promoting new governance means above all promoting a dialogue between the different sectors of the public administration and the agencies responsible for transport, energy, water, housing, food, and health, which do not often coincide, much less interact, on the local and national levels. These circumstances make it difficult to pursue the common objectives of health (Barton and Tsourou 2000), with the risk of conflicts and the dangerous overlapping of fields. It also means guiding a community towards pursuing common well-being, including all civil society—the private sector as well as the media—in public administrative decisions (Kickbusch and Gleicher 2012). As well, it means linking governmental and civil institutions, local, regional, and global communities of interest, creating safe or reliable spaces for diverse interests to interact, and integrating solutions for health problems with solutions for other urban concerns. Finally, it means creating new alliances and making institutions, experts, and citizens responsible for common objectives based on integrated, shared knowledge of the city and its criticalities and needs (D'Onofrio and Trusiani 2017).

This change in mentality should also regard a different model of planning and management. In the book by de Leeuw and Simos mentioned above (2017), Helen Wilding and other scholars, recalling a statement by Kickbusch, maintain that healthy cities faced with complexity need to develop an adaptive learning approach that involves both local and global scales, which necessarily requires a different management style and a different planning model (Wilding et al. 2017). The new management style foresees sharing a healthy city project with civil society, implying continuous negotiation rather than the imposition from above, collegial interactions rather than hierarchical relationships, a holistic approach rather than a sectoral approach (Hancock 1993). The need for integrated knowledge also imposes a revision of traditional cognitive and evaluative models of urban planning, as well as a revision of the traditional hierarchical approach of rationalist planning based on rules and control.

With the city project emerging from Phase VI of the program, the urban planning discipline is basically called to question its values, goals, and objectives again, as well as the strategies to implement to reorganize the space.

Without referring to complex assessment processes, it is evident today that traditional social policies for the city—for example, education, infancy, sports, etc.—show good integration with health-related themes (this is the case of Belfast reported in Appendix 1). On the contrary, policies dealing with improving the living environment of city inhabitants seems less attentive to these themes. Housing, urban planning, and transport are the poor relatives of health, despite innovative—and positive—experiences in some European cities, as some study cases reported in the two Appendices show. As an example, the development of biking and walking present in many policies in many European cities is mainly designed as a factor to reduce atmospheric pollution and traffic congestion and only rarely as a tool to encourage physical activity.

What emerges from the most successful experiences is that the links between local players are a determining factor in realizing a successful project. A successful project is based on the cooperation between city services, interaction with local stakeholders, citizens' associations, etc. It is not by accident, for example, that in the good practice of Rennes presented in Appendix 2, the "*Santé, habitat, environnement*" commission of the inter-sector committee "*Ville Santè*", involves other figures in addition to public health professionals. The commission includes a technician from the urban community-planning department, the Air Breizh Association, which is responsible for monitoring air quality on a regional level, and various teachers/researchers from the School of Public Health (*Ecole des hautes études en santé publique*, EHESP). This is also the case in Belfast, where work on the healthy city is even coordinated by a voluntary sector partnership that serves as a platform for inter-sector collaboration since the local government has limited responsibilities in terms of health (Appendix 1). The City of Bologna, for its part, is pursuing a mode of "doing urban planning" on a local scale, which has become consolidated over time and is based on listening and the community's participation and involvement. The City of Bologna has developed a form of city planning and management based on integrating the various sectors of public administration and being sensitive to the needs of citizens in all age and social categories (Appendix 1).

Despite the difficulties, the Healthy City project is making inroads into a new "idea" of city, a new means of organizing functions in space, composing the urban form, organizing the city's relationship with the environment and the landscape—in effect, a new model of urban planning.

In 2012, the Lancet Commission, in its report "Shaping cities for health: complexity and the planning of urban environments in the 21st century", prefigures a real challenge for HUP, that is, developing a planning model amid complexity. It is a model, therefore that is no longer activated as a well-structured sequential process, but rather through a series of events occurring over time. In this mode, the public decision-maker is no longer a manager, but rather a participant in the city-building process and a facilitator capable of promoting a dialogue among the different actors, suggesting possible solutions to the problems (Rydin et al. 2012).

This new planning model raises important questions regarding the nature of the development strategy to be implemented. This does not mean speculating about an urban plan that anticipates all future changes, but rather a plan constructed through incremental, experimental attempts that identify and promote a wide range of interventions to favour the health and well-being of city inhabitants. These interventions are assessed and monitored over time so that one can learn from them both in the case of favourable results and in the case of possible failures.

References

Barton, H., & Tsourou, C. (2000). *Healthy urban planning. A WHO guide to planning for people*. London and New York: Spon Press.

Barton, H., & Grant, M. (2011). Urban planning for healthy cities: A review of the progress of the European healthy cities programme. *Journal of Urban Health, 90*(Suppl 1), 129–141. https://doi.org/10.1007/s11524-011-9649-3.

Corburn, J. (2009). *Toward the healthy city: People, places and the politics of planning*. Cambridge: MIT Press.

DAC & CITIES (2014). *Health and sustainable cities*. http://www.dac.dk/en/dac-cities/sustainable-cities/background-articles/health-and-sustainable-cities/. Accessed June 10, 2017.

de Leeuw, E. (2011). Do healthy cities work? A logic of method for assessing impact and outcomes of healthy cities. *Journal of Urban Health, 89*(2), 217–231. https://doi.org/10.1007/s11524-011-9617-y.

de Leeuw, E., & Skovgaard, T. (2005). Utility-driven evidence for healthy cities: Problems with evidence generation and application. *Social Science and Medicine, 61*(6), 1331–1341. https://doi.org/10.1016/j.socscimed.2005.01.028.

de Leeuw, E., Green, G., Tsouros, A., & Devlin, J. (2015). Healthy cities phase v evaluation: Further synthesizing realism. *Health Promotion International, 30*(suppl 1), i118–i125. https://doi.org/10.1093/heapro/dav047.

de Leeuw, E., & Green, G. (2017). The logic of method for evaluating healthy cities. In E. de Leeuw & J. Simos (Eds.), *Healthy cities. The theory, policy, and practice of value-based urban planning* (pp. 463–487). New York: Springer.

D'Onofrio, R., & Trusiani, E. (2017). *Città, salute e benessere*. Milano: FAngeli Editore.

Goumans, M., & Springett, J. (1997). From projects to policy: 'Healthy cities' as a mechanism for policy change for health? *Health Promotion International, 12*(4), 311–322. https://doi.org/10.1093/heapro/12.4.311.

Hancock, T. (1993). Health, human development and the community ecosystem: Three ecological models. *Health Promotion International, 8,* 41–47.

Kickbusch, I. (1999). Global + local = glocal public health. *Journal of Epidemiology and Community Health, 53,* 451–452.

Kickbusch, I. (Lead Author) (2010). Triggering debate, White Paper: The Food System—A prism of present and future challenges for health promotion and sustainable development. Health Promotion Switzerland. http://www.ilonakickbusch.com/kickbusch-wAssets/docs/White-Paper—The-Food-System.pdf. Accessed June 10, 2017.

Kickbusch, I., & Gleicher, D. (2012). *Governance for health in the 21st century.* www.euro.who.int/__/RC62BD01-Governance-for-Health-W. Accessed June 10, 2017.

Pilkington, P., Grant, M., & Orme, J. (2008). Promoting integration of the health and built environment agenda through a workforce development initiative. *Public Health, 122*(6), 545–551.

Rydin, Y., Bleahu, A., Davies, M., Davila, J. D., Friel, S., De Grandis, G., et al. (2012). Shaping cities for health: Complexity and the planning of urban environments in the 21st century. *The Lancet, 379*(9831), 2079–2108. https://doi.org/10.1016/s0140-6736(12)60435-8.

Roué-Le Gall, A., Le Gall, J., Potelon, JL., & Cuzin, Y. (2014). *Agir pour un urbanisme favorable à la santé, concepts & outils; Guide EHESP/DGS*. www.ehesp.fr/wp-content/…/2014/09/guide-agir-urbanisme-sante-2014-v2-opt.pdf. Accessed June 1, 2017.

Takano, T. (2003). *Healthy cities and urban policy research.* Abingdon: Routledge.

Taylor, M. (2010). *The Healthy Cities Movement.* (Working paper for the Lancet Commission on Healthy Cities). MSc Urban Studies, University College London. https://www.ucl.ac.uk/healthy-cities/outputs/Working_Paper.

Tsouros, A. D. (2015). Twenty-seven years of the WHO European healthy cities movement: A sustainable movement for change and innovation at the local level. *Health Promotion International, 30* (suppl_1): i3–i7. https://doi.org/10.1093/heapro/dav046.

Tsouros, A. D. (2017). Healthy cities: A political project designed to change how cities understand and deal with health. In E. de Leeuw, J. Simos (Eds.), *Healthy cities. The theory, policy, and practice of value-based urban planning* (pp. 488–504). New York: Springer.

UN. (1993). *Agenda 21: Programme of action towards sustainable development*. New York, USA: United Nations Publications.

WHO Regional Office for Europe. (2013a). *Health 2020. A European policy framework and strategy for the 21st century.* http://www.euro.who.int/en/health-topics/health-policy/health-2020-the-european-policy-for-health-and-well- being/publications/2013/health-2020.-a-european-policy-framework-and-strategy-for-the-21st-century-2013.

WHO Regional Office for Europe. (2013b). *Phase VI (2014) of the WHO european healthy cities network: Goals and requirements*. http://www.euro.who.int/__data/assets/pdf_file/0017/244403/Phase-VI-20142018-of-the-WHO-European-Healthy-Cities-Network-goals-and-requirements-Eng.pdf?ua=1.

WHO Regional Office for Europe. (2014). *Healthy cities. Promoting health and equity—Evidence for local policy an practice. Summary evaluation of phase V of WHO European healthy cities network*. In www.euro.who.int.

Wilding, H., Gould, R., Julia Taylor, J., Sabouraud, A., Saraux-Salaün, P., & Simos, J. (2017). Healthy cities in Europe: Structured, unique, and thoughtful. In: *Healthy cities. The theory, policy, and practice of value-based urban planning* (pp. 241–293). New York: Springer.

Chapter 3
Criteria of Healthfulness in Urban Environments: From a Theoretical Debate to Some Early Experiments

Abstract Policies regarding public health, the environment, and urban planning present in many European countries today are still mostly partitioned, as are the economic resources made available by the central governments. It is necessary to change the paradigm and integrate all the different aspects of health and quality of life of residents and city users to define the physical space of the city and ensure that financial resources from different sources can be integrated. To do this, it is necessary to understand the factors that influence an individual's state of health and, more broadly, the health of a community or population, and how urban space relates to them. Briefly, the so-called "determinants of health" need to be understood. Experimentation should be considered as a necessary action to increase awareness of the relationships between urban planning and determinants of health, along with evaluations of the impact of urban plans and projects on health and the quality of life in cities. Experimentation in the URBACT II PIC Program "Building Healthy Communities" has moved along these lines.

Keywords Public health · Healthfulness · Urban environment Determinants of health · Theoretical debate · Urban experimentation

With regard to health, there is a basic contradiction common to many prevention policies that have been implemented by European cities. These affect individuals, who are responsible for their state of health and behaviour. In contrast, the WHO's philosophy focuses on the set of factors of health that affect the urban population. It is necessary to overcome this contradiction based on a unitary understanding of the theme of health and well-being with the awareness that the health of a population depends not only on the quality of health assistance offered to the individual, but above all on the conditions of living, that is, the social, environmental, and economic conditions.

Despite the evidence of these circumstances, policies regarding public health, the environment, and urban planning present in many European countries today are still mostly partitioned, as are the economic resources made available by the central governments. It is necessary to make a change in paradigm and integrate all the different aspects of health and quality of life of residents and city users to define the

R. D'Onofrio and E. Trusiani, *Urban Planning for Healthy European Cities*, SpringerBriefs in Geography, https://doi.org/10.1007/978-3-319-71144-7_3

physical space of the city and ensure that financial resources from different sources can be integrated.

To do this, it is necessary to understand the factors that influence an individual's state of health and, more broadly, of a community or population, and how urban space relates to them. Briefly, it is necessary to understand the so-called "determinants of health".

The cultural debate on this theme began in the 1970s following the publication in 1974 of the Lalonde Report,[1] which proposed a global approach to health based on four main families of determinants: human biology, environment, lifestyle, organization, and health assistance.

Today, many illnesses that afflict the developed Western world are considered "lifestyle" diseases. Heart disease, stroke, obesity, diabetes, and even some types of cancer have been attributed to our habits and/or conditions of existence. An impressive body of research in the last thirty years has revealed the importance of socioeconomics in determining the state of our health. This growing interest has favoured the spread of other models representing the determinants of health. These include the Dahlgren-Whitehead model (1991)[2] and, more recently, the model developed by the Quebec Ministry of Health and Social Services (MSSS 2010).[3] The interest in this theme shown on the worldwide level was widely adopted by the WHO's Commission on Social Determinants of Health (CSDH), instituted in 2005 under the aegis of Michael Marmot. The charge given to the commission was to identify the ways of reducing health disparities and define what would be necessary to make progress in this direction on the worldwide level.

---

[1]In 1974 the Working Paper by the Canadian Federal Government, "A new perspective on the health of Canadians" (the Lalonde Report), stated that modifying the lifestyle and physical and social environment would probably lead to improving health more than investing money in existing health services.

[2]This model focuses attention on inequalities in matters of health. In fact, significant inequalities regarding the state of health of the population around the world cannot be explained by individual or genetic differences. Dahlgren and Whitehead represent this concept through a graphical representation of the determinants of health organized into four levels of influence around the factors of human biology. The model is expressed in a series of concentric layers, each corresponding to a different level of influence. At the centre there is the individual, with his or her biological characteristics: sex, age, genetic history, i.e., the determinants of health that cannot be modified. The modifiable determinants, those that can be corrected and changed, are organized in layers from the inside to the outside: individual lifestyles, social and community networks, living and working environment, the political, social, economic, and cultural context. This is a conceptual model that both reflects the European culture of the welfare state based on "right to health" and adopts the "multi-sector" vision of the protection of health contained in the Declaration of Alma-Ata.

[3]The conceptual framework of health and its determinants according to the MSSS model (2010) outlines a more detailed vision with respect to the Dahlgren-Whitehead model. It describes health (overall, physical and mental, psychosocial) as a variable that is susceptible to change over time and according to the place. It identifies four large families of determinants (global context, systems, living environment, and individual characteristics) as well as subfamilies (biological and genetic characteristics, personal and social abilities, family, school, and childcare) that influence the state of health of the population. The model also allows the importance of territorial planning choices for public health to be considered.

The final report was published in 2008 with the title "Closing the gap in a generation. Health equity through action on the social determinants of health".[4] The document focused not only on factors influencing the state of health of individuals and communities (determinants of health), but also those involved in the unequal distribution of the quality of health within the population (determinants of health inequalities). The factors were traced to the political and socioeconomic context, socioeconomic position, working and living conditions, psychosocial factors, social cohesion, individual behaviours and biological aspects, and the health system.

In 2013, the WHO's report "Review of determinants and the health divide in the WHO European Region: executive summary" investigated these aspects further, with the aim of providing recommendations to reduce inequalities in terms of health (WHO Regional Office for Europe, UN City 2013). These recommendations revolve around four cardinal themes:

- throughout life (in family building, childhood and the work place, and for women and older people);
- in wider society (action for social protection, on gender, in local communities and against social exclusion);
- on the macro-level context (on social expenditure and for sustainable development); and
- on systems (to improve governance, to address health priorities and to measure and report on progress against inequities in health).

With reference to the themes more specifically dealt with in this book, in reference to the cardinal theme "macro-level context", one recommendation regards the link "sustainable development and health". Long-term planning and the protection of the interests of future generations are suggested, identifying connections between environmental, social, and economic factors and their centrality in strategies and practice.

For the cardinal theme "systems", the report instead recommends improving the governance of the social determinants of health and health equalities/inequalities. Greater coherence is therefore required in actions on all levels of government—transnational, national, regional, and local—and across all sectors of stakeholders—public and private entities and volunteers.

Even before the WHO's 2013 report, Barton and Grant were moving in this direction in 2006, gathering the relationships between the determinants of health and territorial planning (Barton and Grant 2006). They crossed Dahlgren and Whitehead's model of determinants of health (1991) with the theory of urban ecosystems, developing a common framework of reference to identify the leverage that each local actor has in promoting urban planning favourable to health. Official reports from international entities and research activities illustrate how territorial planning has for some time been considered a powerful tool for creating living environments

[4]CSDH (2008). Closing the gap in a generation: health equity through action on the social determinants of health. Final Report of the commission on Social Determinants of Health. Geneva: World Health Organization.

that are increasingly favourable for health. Despite a strong dedication (at least on paper) to introducing the theme of health in developing programs, plans, and projects, in practice this occurs only rarely and with great difficulty.

Recognition of the dimension of health and well-being in urban policies is essentially tied to environmental determinants (quality of the physical environment, air, water, noise, land, etc.) and according to an approach based on risk, omitting many determinants of health related to the social or economic areas or more broadly to the living environment (Harpet and Roué le Gall 2013).

As highlighted above, in order to bridge this gap, better collaboration between and integration of the sectors of public health, environmental planning, and urban planning will inevitably be found, as also evinced by the report by the Lancet Commission, "Shaping Cities for Health Commission" (Rydin et al. 2012).

Two themes of interest for this book can be extrapolated from the Lancet report: the value of experimentation as a necessary action to increase awareness of the relationships between urban planning and determinants of health; and the role of evaluating the impact of urban plans and projects on health and the quality of life in cities.

Experimentation in the URBACT II PIC Program "Building Healthy Communities" (2011) has moved along these two lines. This is a program promoted by a network of ten European cities in seven European Member States (Amaroussion, Bacău, Baia Mare, Barnsley, Belfast, Lecce, Lidingö, Łódź, Madrid, Turin) that decided to work together from 2008 to 2011 in order to capitalize on knowledge and practices related to urban factors that influence health. The goal is also to create opportunities for cities to build and implement policies to improve the quality of life and health of their citizens. The ten partner cities in BHC aimed to design Local Action Plans (LAPs) in which health and the quality of life can be considered principal objectives and key words for orienting development.

All of this comes at a time when the economic and financial crisis is strongly affecting European economies, forcing cities to reduce their available budgets, establish new priorities, redefine development strategies, and refocus the challenges to address, planning interventions according to long time horizons. In this situation, the ability to assess policies and monitor them over time assumes great importance.

The objective of the BHC project was to provide a sort of guide to the city, a document that, starting from local specifics and needs, manages to assist cities in the process of building healthy communities. Ten very different guides have been created because the needs, approaches, political and social culture are different from city to city. Despite this, such diversity constitutes an advantage because the process of learning from the different experiences has become a line supporting the project. In comparing the experiences, innovative means have emerged, such as: how to assess and monitor health in the city; the different policies to implement; and locally funded opportunities to realize projects in favour of health and urban quality. The objective of "health in all policies" was crossed with current processes of urban regeneration in the different local contexts. This juxtaposition of theory and practice has led to the development and testing of indicators and criteria to evaluate plans for urban regeneration and plans for city development with the final objective of achieving "good health" and sustainable development.

The set of indicators chosen regard three main areas: economic development, cultural and social cohesion, and environmental regeneration.

The "control list" developed to assess the different proposals for regeneration included direct references to factors of health but also the characteristics of the communities and neighbourhoods, such as the level of education, access to services, and the presence of green areas and transport. These indicators constitute a framework of reference useful for defining the areas of intervention, the different levels of responsibility, and cooperation to develop an integrated approach to urban regeneration. The type of assessment thus outlined has been shown to be very useful for the cities in terms of identifying priority questions to answer through practice.

Many cities have included the theme of health and well-being within the plans and projects for urban regeneration, which are in many cases already underway. This is the case, for example, in Turin, with its project for the Barriera di Milano neighbourhood, inserted within a program of integrated development (*Programma di Sviluppo Integrato*, PISU).

In Madrid, the BHC project was inserted within a strategy to develop the historical centre and aimed to make Embajadores, a central part of the city, a more pleasant place through measures for social integration, sustainable development, culture, modernization, and safety. The project established three main objectives: discover new modes of experiencing/living in the city; facilitate communication between two large green areas in the historical centre, and promote a healthier lifestyle by improving pedestrian infrastructures. The project used different indicators to monitor the quality of the places from the point of view of citizen health. This way of assessing projects is new for Madrid, and has aided the construction of a common language that accompanies the project and helps the project itself to be redesigned.

A second project in Madrid, which was the object of the LAP, regards the Madrid Río project to create a park along the Manzanares River in order to create a connection between the city centre and the innermost area of the urban belt. The LAP developed three themes:

- design a working methodology and an assessment tool;
- apply this methodology to a study area and verify it;
- select pilot actions to develop the proposal.

As developed, the methodology allowed the risks for health to be identified, which could be helpful in analyzing the effects and impacts of public interventions and in reflecting on the concept of vulnerability in a holistic way. As well, it also allowed measures for monitoring and control to be implemented with the direct involvement of the urban community.

The experiences accrued by the different cities in the URBACT Network also highlighted the difficulties related to such an inclusive approach. The first regards the involvement of management authorities in creating the LAP. It was seen that regional and national authorities had often already decided everything, for which the role of the local community was limited. In addition, it emerged that on the local level there is an urgent need to promote integrated interventions to improve the quality of life of citizens and establish works for prevention. In this sense, the need to integrate resources from different sources is necessary (community, national, local, private financing) and to promote attention for health even in existing programs in order to interpret, adapt, and brainstorm different solutions.

The conclusive report of this program does not request an increase in the skills of the city in terms of health, but rather highlights the need to improve the knowledge of citizens, politicians, and experts in matters of quality of life and health, to better define the development policies to implement. The affirmation therefore holds that health is a local political choice and that it is not obligatory for cities to put health at the centre of their strategies; if they do so, it is to respond to the need to improve the life of their citizens.

The results of the BHC program can be summarized in three areas of interest for our research:

- a focus on the theme of assessing health in urban policies. This attention has implied a "toolbox" of indicators that allows the level of health and quality of life in European cities to be measured;
- understanding how cities can positively influence the lifestyle of their citizens;
- better using the available resources, even reworking current projects for urban regeneration, incorporating objectives related to health and the quality of life.

Experience amassed within this European program moves precisely in the direction indicated by the recommendations for Phase VI of the Healthy Cities movement. The recommendations express the need to revisit the approach followed up to now because the theme of health requires specific working methods capable of interacting and understanding the context and the impact of the context on policies.

Some of the themes of interest emerging from the experimentation in the URBACT model were used to assess the case studies presented in Appendix 2 related to: (a) the role of assessment in measuring the level of health and quality of life in the selected cities; (b) forms and means of involving the communities and interest holders in formulating plans, projects, and the assessment processes within the HIA; (c) the effective capability of local communities to orient processes to improve the quality of health and urban life and influence decisions; (d) the capacity of local urban planning to involve the different sectors of the public administration (with particular reference to the sector of public health) to formulate common projects and plans; (e) taking effective charge in local plans and projects of the determinants of health (not only environmental, but also social).

References

Barton, H., & Grant, M. (2006). A health map for the local human habitat. *The Journal for the Royal Society for the Promotion of Health, 126*(6), 252–253. https://doi.org/10.1177/1466424006070466.

Building Healthy Community. (2011). *Building healthy communities final report*. http://urbact.eu/bhc-results. In http://urbact.eu/building-healthy-communities. Accessed June 1, 2017.

CSDH. (2008). *Closing the gap in a generation: Health equity through action on the social determinants of health.* Final Report of the Commission on Social Determinants of Health. World Health Organization: Geneva. whqlibdoc.who.int/publications/2008/9789241563703_eng.pdf. Accessed June 1, 2017.

Harpet, C., & Roué Le Gall, A. (2013). Évolution de la prise en compte de l'environnement et de la santé dans le champ de l'urbanisme. *Environnement, Risques et Santé*, *12*, 231–241.

MSSS. (2010). *Plan Stratégique 2010–2015*. http://publications.msss.gouv.qc.ca/msss/document-000295/. Accessed June 1, 2017.

Rydin, Y., Bleahu, A., Davies, M., Davila, J. D., Friel, S., De Grandis, G., et al. (2012). Shaping cities for health: Complexity and the planning of urban environments in the 21st century. *The Lancet, 379*(9831), 2079–2108. https://doi.org/10.1016/S0140-6736(12)60435-8.

WHO Regional Office for Europe, UN City. (2013). *Review of social determinants and the health divide in the WHO European Region: Final report*. www.euro.who.int/__.../Review-of-social-determinants-and-the-health-divide-in-the-W. Accessed June 1, 2017.

Part II
Healthy Urban Planning in Europe

Chapter 4
The Need for New Urban Planning for Healthy Cities: Reorienting Urban Planning Towards Healthy Public Policy

Abstract The role of local public administrations in favoring dialogue among different players—professionals and workers in public health, urban planners, politicians, and the civil community—is fundamental. All these figures can assume common responsibilities in constructing plans and projects for the city, overcoming the gap between different skills, approaches, and languages. What is certain is that the role of the central (national) government in health policies and city planning cannot be ignored: national approaches, laws, and regulations affect local plans and policies. Some experiments made in recent years in Finland, the United Kingdom, Sweden, Denmark, etc., on attempts to integrate urban planning and health have felt the effects of some important innovations on the central level [public-health reforms, national recommendations, guidelines on the Health Impact Assessment (HIA), etc.]. Other countries, such as France and Italy, began to address these themes only a few years ago, and on the national level, legislative references are still lacking. Little experimentation has been made in the field, and the experiments carried out have mostly regarded the application of the HIA to individual plans and projects. At any rate, there is growing interest even in these countries.

Keywords National and local governance · Public-health reforms
National recommendations · Guidelines on the Health Impact Assessment (HIA)
Urban planning and healthy public policy

Adopting the strategy "Health in all Policies" does not mean working exclusively on the availability of health services and promoting healthy lifestyles, but rather acting on the quality of environments and living/working conditions, improving citizens' economic resources, community cohesion, and the provision of quality public services. Local public administrations play an important role in promoting this new idea of protecting health in policies, plans, and projects. In many European countries, local entities are often directly involved in managing strongly decentralized health systems, taking on many responsibilities related to the quality of services and people's living spaces (Progress Consulting 2012). It is on the local scale that health and well-being can be promoted in daily practice, closely intercepting the real needs of citizens (Berkeley and Springett 2006). This is one of the motives why the World Health Organization (WHO) has chosen to "jump over" national governments

R. D'Onofrio and E. Trusiani, *Urban Planning for Healthy European Cities*, SpringerBriefs in Geography, https://doi.org/10.1007/978-3-319-71144-7_4

and communicate directly with cities.[1] The role of local public administrations in favouring dialogue among different players—professionals and workers in public health, urban planners, politicians, and the civil community—is fundamental. All these figures can assume common responsibilities in constructing plans and projects for the city, overcoming the gap between different skills, approaches, and languages, as the experience in Belfast reminds us.

What is certain is that the role of the central (national) government in health policies and city planning cannot be ignored: national approaches, laws, and regulations affect local plans and policies. Some experiments made in recent years in Finland, the United Kingdom, Sweden, Denmark, etc., on attempts to integrate urban planning and health have felt the effects of some important innovations on the central level [public-health reforms, national recommendations, guidelines on the Health Impact Assessment (HIA), etc.].

Other countries, such as France and Italy, began only a few years ago to address these themes, and on the national level, legislative references are still lacking. Little experimentation has been made in the field, and the experiments carried out have mostly regarded the application of the HIA to individual plans and projects. At any rate, there is growing interest even in these countries.

Below, Boxes 1, 2, and 3 present some national experiences in England, Wales and Scotland, and France and Italy. These different examples show how the road to integrate urban planning and health is not always simple; it is the fruit of a cultural, habitual way of addressing complexity in an integrated manner. The level of integration does not depend as much on the system of planning to be addressed as on leadership, dedication, and the skills of the politicians and practitioners involved. The true barriers to integration often lie in organizational and professional silos. This condition requires local administrations, through its agencies and planning services, to play a primary role in favouring the creation of strong partnerships with those responsible for decision-making in terms of health, transport, housing, economic development, etc. However, contributions from the central level of public administration are also necessary. They should work to orient the identification of tools and guidelines to adapt to the local scale.

The road to integration between health and urban planning is anyway a necessary road because it is "convenient" to follow for various evident reasons. The first can be summarized in the statement that "a healthy city is a city that can balance its accounts". We avoid any complex analysis, but based on daily experience of administrators and citizens, it is easy to understand how, for example, the aging of the population in the coming years will lead to exponential growth in the request of some services. As well, cities are largely unprepared to face this emergency, both in economic terms and in the design of living spaces. On the other hand, today we are able to quantify the monetary savings of a city that promotes, for example, the physical activity of its citizens.

---

[1] Among the motives, Goumans and Springett (1997) highlight the lack of recognition by European governments of the role played by social/environmental conditions in determining health and well-being.

"A euro spent by the administration translates into ten Euros not spent for the health service…because physical activity implies overall well-being". These words were expressed some years ago by Simona Arletti, the president of the Italian Healthy Cities Network, in reference to the use of the Heat application (Health Economic Assessment Tool) in Modena. This is a tool promoted by the European Commission and the WHO to assess the economic impact of health. It calculates the savings in Euros produced by an intervention to promote the health of citizens and estimates the reduction of mortality if physical activity becomes a habit and not an exception. A city that addresses these themes in planning and design will not only be a "healthy" city, but will also probably be a city that has public spending under control.

A second reason, which is more strictly disciplinary, regards the opportunity provided by this partnership for a "renewal" in urban planning along the lines of a renewed interest in "living", a new local welfare that can favourably influence the living conditions of city inhabitants. This opportunity is demonstrated in reference to some primary aspects:

- a focus on the integrated, shared knowledge of the city and its needs and criticalities in terms of health and well-being as an opportunity to create alliances and hold institutions, experts in the different disciplines, and citizens responsible;
- the move from a regulatory approach to a need-/performance-based approach in designing/regenerating the city and living spaces. Rethinking spaces, placing the health of people at the centre, can help to move the focus of design from "how to intervene" to "why to intervene", inviting consideration of the real demand that determines particular actions or the construction of urban spaces and their management;
- the long-term adaptability of the projects proposed, which stimulates recourse to voluntary agreements and pacts with citizens to maintain, manage, and decorate living spaces, opening incremental practices in urban regeneration;
- placing the local plan as the preferred reference for the experiences and practices of horizontal, bottom-up subsidiarity, aimed at sharing projects and responsibility in the co-production and management of services (including non-conventional ones), generating spaces and opportunities for social inclusion.

Chapters 6 and 7 and Appendix 2 summarize some experiences in European cities where such disciplinary innovations can be assessed and verified.

4.1 Box 1—England

England is among the countries that have been working the longest to integrate health and urban planning.

Starting in 2010 with the Marmot Review "Reducing health inequalities in England" and the white paper "Healthy Lives, Healthy People", the country has dedicated itself to taking actions to reduce health inequalities among the population, in the conviction of the usefulness of a transverse approach. It has therefore begun a complex task of cultural and evaluative discussion with many actors in public policies and fully involving urban planning.

In a 2013 article entitled "Health-integrated at the local level in England: impediments and opportunities", Carmichael, Barton, et al., assessing the recent experiences in England, observed that the planning system did not contain any specific planning policy guidance or planning policy statement on health (Carmichael et al. 2013). Despite this, the authors could not help but detect the large provision of non-statutory healthy environment guidance in the fields of urban design, sustainable building design, local transport and street design, open spaces, green-spaces, and recreation [for instance: By Design (DETR and CABE 2000); the Code for Sustainable Homes (DCLG 2010); Building for Life (CABE 2008); the Environmental Impact Assessment: A guide to procedures (DCLG 2000; etc.)].

This large availability of tools, which very often hold directional value and are capable of intercepting local specifics, does not make up for the lack of a general orientation. This is reflected in the nature of local experiences, which are very diverse and have uncertain success because they are tied to the capacity/incapacity of the individual administrations to include themes of health within policies and plans.

Starting in 2012, there has been profound legislative and social reform in reference to three legislative texts: the National Planning Policy Framework (NPPF) from 2012; the Health and Social Care Act of 2012; and the Localism Act of 2011.[2] This reform has theorized that urban planning and policies for health can be achieved by broadly expanding the mandate of local authorities. The NPFF in particular suggests three aspects: more joint collaboration between public health and planning; deep involvement in building plans for the development of public health and health organizations to understand and keep in mind the state of health and the needs of local populations; and active community involvement.

[2]The first of these, the National Planning Policy Framework (NPPF) of 2012, fixed the general guidelines in which local administrations should provide urban plans with the expressly declared objective of aiming for healthy cities, assessing needs in terms of well-being, and collaboration with health authorities and their structures.

The second is the Health and Social Care Act of 2012, which transferred the responsibility for public health to local communities starting in 2013 and which also called for the constitution of mixed commissions on health and well-being among national service workers and local entities to strategically program health and social services.

The third, the Localism Act of 2011, conferred more power on local entities and inhabitants, allowing them to redirect even the Neighbourhood Plans in a participatory way. It also introduced a series of innovations influencing the management of health, including social housing and its creation and management.

Important work to publicize the reform's innovations was undertaken starting in 2012 by the Town and Country Planning Association (TCPA),[3] with its project "Reuniting Health with Planning". Through various publications such as "Reuniting Health with Planning Handbook" (2012), "Planning Healthier Places" (2013), and intensive communication activities, the TCPA has promoted a variety of events and debates. These are aimed at informing public opinion, designers, and administrators of how these legislative innovations can lead to significant benefits in terms of health and well-being, and how they can be pursued through local urban planning. More concretely, the TCPA has often provided practical support for local administrations in developing plans and guidelines. This has occurred, for example, with the "Good Practice Guide" (TCPA 2015a) in support of the London Local Plan[4] (City of London 2015). The guide, together with a detailed document for the different neighbourhoods—the "Good Practice Guide for London Boroughs" (TCPA 2015b)—identified a model to make an initial screening in the planning process to highlight the presence of potential risks for health in order to reduce or attenuate the possible impacts. The guide also provided HIA-type indications and suggestions and possible modifications in planning choices, which should then be transferred to the screening activities. Finally, it calls for a careful study of the current state of local plans (Boroughs' Local Plans) and considers the possibility of modifying lower-level tools such as the Supplementary Planning Documents, specific strategies for open spaces, green spaces, infrastructure, urban regeneration, and to guarantee a proactive process to assess the effects on people's health.

[3]This is a charity founded in England in 1899 that promotes knowledge regarding territorial planning and sustainable development.

[4]The London Local Plan is the main urban plan for the City of London. It establishes: the vision for the city for the next 15–20 years; policies in matters of land use, housing policies, transport, urban regeneration, and environmental choices; and the guiding principles for lower-level subordinate planning.

Some years since the reform, the impression of some scholars is that there is still a lot to do, but that, although "…health-integrated planning and appraisal…" has still not been verified, "there are tools which support any local area to get started…" (Barton et al. 2015). Experimentation therefore becomes the only possibility for continuing down this road, integrating the skills of experts in the health sector, developers, and planners. All of this lies within the scope of understanding different languages and expectations, with the goal of obtaining concrete advantages from the discussion. In addition, the hope is that the NPFF facilitates the debate on the national scale, in the awareness that integration is not a problem of planning tools as much as it is of experimentation.

Growing interest in "practice", even on behalf of the state, emerged in 2014 when the British National Health Service (NHS), in its Five Year Forward View (NHS 2014), launched the programme "Healthy New Towns" to improve the health of the population and integrate health and assistance services. The ambition of this programme, which was also developed to respond to the requests of more than 200,000 households in England each year, is to go beyond existing good practice in terms of health and well-being. In fact, the goal of this initiative is to promote the development of new, creative approaches to create new cities and to renew some degraded urban areas; offer the opportunity to design modern services with fewer restrictions; and integrate not only health and social assistance in design, but also other public services such as education, accessible housing, etc. The goal of the programme, which has published a call and selected ten cities for experimentation,[5] was also to promote a closer collaboration among local authorities, designers, developers, and the NHS on the national level.

In addition, the usefulness of a closer collaboration between central authorities (the NHS) and local authorities is being tested with the institution of the Health and Wellbeing Board following the emission of "Health and Social Care" in 2012. These statutory bodies have the objective of improving integration between practitioners in local health care, social care, public health, and related public services. They are also responsible for local leadership to reduce health inequalities, favour the possibility of working closer with themes such as urban planning and urban regeneration, housing, free time, transport, and mobility, and provide assistance in identifying local priorities that are coherent with national strategies.

[5]Whitehill and Bordon, Hampshire; Cranbrook, Devon; Darlington; Barking Riverside; Whyndyke Farm in Fylde, Lancashire; Halton Lea, Runcorn; Bicester, Oxon; Northstowe, Cambridgeshire; Ebbsfleet Garden City, Kent; Barton Park, Oxford.

4.2 Box 2—Wales and Scotland

In addition to England, Wales and Scotland have recently developed new proposals to implement the themes of health and well-being in urban planning. In Wales, in particular, three legislative texts have been issued: the Planning (Wales) Act of 2015, the Well-Being of Future Generations (Wales) Act 2015, and the Environment (Wales) Act of 2016, the last of which promotes sustainable development, the correct use of natural resources, and guarantees the well-being of future generations.

Planning for health and well-being is not a novelty for the country. The three laws mentioned above, however, clearly involve the planning profession, with input from public health professionals, to improve health and well-being. To facilitate this collaboration, the Wales Health Impact Assessment Support Unit (WHIASU) and the Town and Country Planning Association (TCPA) published "Planning for Better Health and Wellbeing—A briefing on integrating planning and public health for practitioners working in local planning authorities and health organisations in Wales" in 2016. This document favours the planning and public health sectors working together to support health improvement and strengthen planning sector plans and decisions.

In the UK panorama, Scotland acted as a pioneer in 2008 with the initiative "Good Places, Better Health". This was a governance strategy that introduced a new approach to themes of health with respect to living environments. The approach goes beyond traditional attention for risks to health inherent in the urban environment to actively contribute to creating "health-friendly" places. Experimentation was carried out from 2008 to 2001 and considered children's health to address aspects such as obesity and asthma, accidental injury, mental health, and well-being. More generally, the experimentation was aimed at searching for new, more effective approaches to plan and design the urban environment, with the goal of contributing to the creation of places that favour people's health and well-being and to reduce health inequalities (The Scottish Government 2008). Many associations, organizations, and governmental agencies, including NHS Health Scotland and Architecture and Design Scotland, have been initiated to promote and support this project.

In the meantime, the legislative approach has changed, producing important innovations in the content of urban planning and within the entire system. Through the National Planning Framework 3 (NPF3) and the Scottish Planning Policy (SPP) of 2014, priorities were identified in the system of national planning and sustainable development. In particular, the SPP introduced a series of principles to favour forms of sustainable development that are tightly integrated with the concept of health (The Scottish Government 2014), which include: supporting mitigation and adaptation to contrast the effects of climate change, including the risk of flooding; favouring the improvement of health and well-being; offering opportunities for social interaction and physical activity, including sport and recreation; protecting, enhancing, and promoting access to the natural heritage, the green network, the landscape, and the environment in general; protecting the amenities of places (both new and existing); and considering the effects of development on the water, air, and land. The SPP is accompanied by the following documents: "Creating Places", which contains policies and directions for architecture and design; "Designing Street",[6] which contains principles and directions to design roadways, resurface and maintain existing streets; and the "Circulars", which deal with procedural content. The contents of the SPP are to be implemented in Local Development Plans (SDP).

4.3 Box 3—France and Italy

In France and Italy, the road to integration still has a long way to go. The themes of health and well-being often enter in city planning and design only through isolated experiments promoted by European Union programmes or due to preliminary HIAs, which represent an attempt to institutionalize and regulate this relationship.

In France, various national reforms in matters of the environmental assessment of projects ("Project" Decree no. 2011-2019) and plans [Decree no. 2012-616—for the environmental assessment of some plans/schemes/programmes—and Decree no. 2012-995—related to assessing urban plans (PLUi, PLU, SCoT[7])] make it necessary to consult the Health Authorities (*Autorité de santé*, AS) regarding their possible impacts on the environment and on health. Following this, increasing numbers of urban policies and projects are subject to evaluation.

---

[6]See http://www.scotland.gov.uk/Publications/2010/03/22120652/0.

[7]The PLUi is the *Plan Local d'Urbanisme intercomunale* [Inter-Municipal Local Urban Plan], the PLU is the *Plan local d'Urbanisme* [Local Urban Plan], and SCoT is the *Schéma de Cohérence Territoriale* [Territorial Coherence Plan] (regarding many municipalities).

The development of the guide "Agir pour un urbanisme favorable à la santé"[8] [Actions for Health-Friendly Urban Planning] from 2014 (EHESP/DGS 2016) constitutes valid support in this direction. It has allowed different themes to be developed for reflection and the creation of an assessment matrix useful for assessing the PLU and (if appropriate) the SCoTs. The assessment matrix, which is based on nine determinants of health,[9] is integrated with another summary document that investigates the project by considering five axes that respond to five basic questions about sustainable urban planning. Does the project contribute to improving the economic and social environment? Does the project eliminate or reduce social and environmental inequalities? Does the project relieve conflicts and can it overcome and favour the development of common synergies? Does the project promote healthy lifestyles and physical rather than sedentary activity? Does the project produce pollution that puts certain segments of the population at risk?
The HIA does not aim to replace existing tools, but rather to provide a framework of experiences to assist decision-making. Today it is a voluntary approach that complements the strategic environmental assessment (SEA).
In Italy, the lack of a national HIA framework of reference up to 2016 means it has not become widespread except on the level of experimental activities within the Regions. In July 2016, the emission of the HIA guidelines by the Ministry of Health served as an initial response to recommendations made by the European Commission (COM/2002/0276). Despite this, the country counts multiple regulatory and applicative deficiencies with respect to the European recommendations. The HIA is obligatory only for particular projects (Law 221/2015), such as oil refineries, gasification systems, heating systems with a power greater than 300 MW, etc. In addition, Italian legislation, while requiring an assessment of the health component in the environmental impact assessment (EIA) and the SEA according to Law 152/2006, does not provide directions, methods, or tools to give indications to the proponents and assessors in charge of analyzing and evaluating the "health component" in the area of plans and projects. In constructing the EIA and SEA, the guidelines introduce participation of the population and equity as determining factors to ensure the acceptance of the results of the HIA and, therefore, its effectiveness.

---

[8]The General Director for Health asked the School of Health (*Ecole des hautes études en santé publique*, EHESP) to develop a tool to analyze the impact on health of urban projects and to initiate research and skills on ways to promote health in the field of urban planning.

[9]The determinants are: (1) outdoor air quality; (2) water resource quality and management; (3) quality of land and subsoil use; (4) quality of the noise environment; (5) waste management; (6) radiation management; (7) adaptation to climate change; (8) mobility, transport, and access to facilities and services; and (9) housing and living environments.

Thus there is still much work to do to integrate the HIA in urban planning. However, some experimentation should be recalled, such as what was conducted within the Emilia-Romagna Regional Prevention Plan entitled "*Linee Guida per la definizione dei contenuti igienico sanitari degli strumenti di pianificazione territoriale della Regione Emilia Romagna*" [Guidelines for defining hygienic/health content in territorial planning tools for the Emilia Romagna Region] drafted by the regional health service (Regione Emilia Romagna 2010). These guidelines constitute an expansion of the Regional Urban-Planning Law (*Legge Urbanistica Regionale*, Regional Law 20/2000). With reference to the large systems in the anthropized environment (various networks, settlement systems, service networks, rural town centres, etc.), objectives were identified for urban planning tied to the protection and improvement of health and the living and working environment, and directions and actions aimed at pursuing these objectives. The project of the guidelines, published in 2010, was then set aside due to the lack of sharing by decision-makers, in that it was deemed to weigh down the procedure to approve urban-planning tools.

References

Barton, H., Thompson, S., Burgess, S., & Grant, M. (2015). *The Routledge handbook of planning for health and well-being*. Routledge.

Berkeley, D., & Springett, J. (2006). From rhetoric to reality: Barriers faced by health for all initiatives. *Social Science and Medicine, 63*(1), 179–188.

CABE (2008). *Building for Life*. Delivering great places to live: 20 questions you need to answer. London: CABE.

Carmichael, L., Barton, H., Gray, S., & Lease, H. (2013). Health-integrated planning at the local level in England: Impediments and opportunities. *Land Use Policy, 31,* 259–266. http://dx.doi.org/10.1016/j.landusepol.2012.07.008. Accessed June 6, 2017.

City of London. (2015). *Local plan.* https://www.cityoflondon.gov.uk/services/environment-and-planning/planning/planning-policy/local-plan/Pages/default.aspx. Accessed June 6, 2017.

DCLG (Department for Communities and Local Government) (2000). *Environmental Impact Assessment: A Guide to Procedures*. http://webarchive.nationalarchives.gov.uk/20120919132719/www.communities.gov.uk/documents/planningandbuilding/pdf/157989.pdf

DCLG (Department for Communities and Local Government) (2010). *Code for Sustainable Homes*. Technical Guide. https://www.gov.uk/government/uploads/.../code_for_sustainable_homes_techguide.pdf. Accessed February14, 2018.

DETR and CABE (2000). *By Design-urban design in the planning system: Towards better practice*. London: Thomas Telford.

EHESP/DGS. (2016). *Agir pour un urbanisme favorable à la santé: Outil d'aide à l'analyse des Plans locaux d'urbanisme au regard des enjeux de santé*. http://www.social-sante.gouv.fr/…/ehesp_dgs_outil_d_aide_analyse_des_plu_enjeux_de_sante.pdf. Accessed June 6, 2017.

Goumans, M., & Springett, J. (1997). From projects to policy: Healthy Cities as a mechanism for policy change for health? *Health Promotion International, 12*(4), 314.

NHS. (2014). Five year forward view. https://www.england.nhs.uk/publication/nhs-five-year-forward-view/. Accessed June 6, 2017.

Progress Consulting S.r.l e Living Prospects Ltd. (2012). La gestione dei sistemi sanitari negli Stati membri dell'UE. Il ruolo degli enti locali e regionali, Unione Europea.

Regione Emilia Romagna. (2010). *Linee Guida per la definizione dei contenuti igienico-sanitari degli strumenti di pianificazione territoriale*. Testo disponibile in: http://salute.regione.emilia-romagna.it/prp/doc/atti-di-convegno-e-formazio-ni/-Linee_Guida.pdf/view.

TCPA, et al. (2015a). *Public health in planning-good practice guide.* https://www.housinglin.org.uk/_assets/Resources/Housing/OtherOrganisation/TCPA_Public_Health_in_Planning_Good_Practice_Guide.pdf. Accessed June 6, 2017.

TCPA, et al. (2015b). *Public health in planning-good practice guide for London boroughs.* http://tcpa.brix.fatbeehive.com/data/files/Health_and_planning/2015_health/Submit_-_TCPA_FINAL_Policy_Review.pdf. Accessed June 6, 2017.

The Scottish Government (2008). *Good Places, Better Health–A New Approach to Better Health and Environment in Scotland*. https://www.scotland.gov.uk/Topics/Health/Healthy-Living/Good-Places-BetterHealth. Accessed February 14, 2018.

The Scottish Government (2014). *Scottish Planning Policy*. http://www.gov.scot/Resource/0045/00453827.pdf. Accessed February 14, 2018.

Chapter 5
Health Promotion and Urban Sustainability: A Perspective on Duality

Abstract The correlation between human health and sustainability/climate change is the fruit of more than twenty years of research activities in Europe. Health creates the conditions for sustainability while simultaneously being conditioned by it, just as sustainability, intended as environmental, economic, and social sustainability, creates and is conditioned by human health. The two concepts, in theory as well as in practice, cannot be separated, but should be understood as interdependent. This means that strategies oriented towards sustainable development should be correlated with strategies to promote health and vice versa. In this sense, experiences within Europe (London, Barcelona, Copenhagen, Malmö, Rotterdam, and Turin, to name a few) and the rest of the world (Boston, Jakarta, Medellin, New Orleans, New York) constitute an interesting record for extrapolation with respect to some keywords running in the direction of health and quality of life in cities: environmental and social safety, public spaces and inclusive cities, and meaningful design references, on both large and small scales.

Keywords Health and quality of life · Urban sustainability · Adaptation to climate change · Flexibility of urban planning tools · Environmental and social safety

"...The good health of all its citizens is one of the most effective markers of any city's sustainable development". Thus maintains the report "Health as the Pulse of the New Urban Agenda" (WHO 2016) to underline how health now plays a central role in supporting actions to carry out and achieve objectives in the New Urban Agenda promoted by the United Nations Habitat III Conference in October 2016 in Quito, Ecuador. In particular, the report highlights the need for political decision-makers to look through the "lens of health" in order to fully assess risks and opportunities and measure the effects of their programmes. It also provides the opportunity for a greater inter-sector cooperation to protect and improve the health of urban populations undergoing rapid growth throughout the world.

The verification of a deep relationship between sustainability and health constitutes the success of a debate that originated at the Ottawa Conference of 1986 and in

R. D'Onofrio and E. Trusiani, *Urban Planning for Healthy European Cities*, SpringerBriefs in Geography, https://doi.org/10.1007/978-3-319-71144-7_5

the Brundtland Report entitled "Our Common Future" (1987), and which has been developed in recent decades.

At the beginning it was the concept of "environmental health", comprising the aspects of health and illnesses determined by the environment, to be related to themes of development sustainability (WHO 1999). Starting in 1990, the WHO's approach has expanded to include the social and mental conditions that affect health. The WHO has therefore defined "environmental health" as "…those aspects of human health, including quality of life, that are determined by physical, chemical, biological, social and psychosocial factors in the environment. It also includes aspects to correct and prevent "…those factors in the environment that can potentially affect adversely the health of present and future generations" (WHO 1993).

This approach implies that:

- improvements and solutions to problems of health should be found in the elimination of physical, chemical, biological risks to the environment and in improving social conditions;
- to be defined as "sustainable", development should satisfy essential human needs such as work, food, energy, water, hygiene/health services, and health assistance (World Commission on Environment and Development (WCED) 1987).

Health and sustainable development are therefore seen as important conditions in a reciprocal definition and specification. This reciprocity is precisely the starting point to try to add further meaning to this relationship.

Sustainable development is a process that implies a change in paradigm in how development is understood: from development based on inequality and the excessive exploitation of natural resources and environmental services, to development that asks for new forms of compassion, responsibility, and solidarity (Shiva 2005).

For its part, the concept of *health* as developed by the WHO goes beyond the concept of "absence of disease". Health is part of the dynamics of social organization, lifestyles, and models of consumption, and is influenced by the biophysical environment. Therefore, in brief, human health is determined by complex relationships between the social and economic systems, the biophysical environment, and individual characteristics of the person (Kickbusch 2010).

The interpretation of the two concepts evinces their duality: health creates the conditions for sustainability while simultaneously being conditioned by it, just as sustainability, intended as environmental, economic, and social sustainability creates and is conditioned by human health. The two concepts, in theory as well as in practice, cannot be separated, but should be understood as interdependent. This means that strategies oriented towards sustainable development should be correlated with strategies to promote health and vice versa (Pedersen et al. 2015).

This close interrelationship should be noted when addressing the effects of climate change on human health. The report "Health as the Pulse of the New Urban Agenda" introduces the theme of "urban health resilience" as the need to reinforce health systems to be able to anticipate and respond to health crises. The report underlines the need to propose resilient urban policies to address themes related to health, to develop common assessments (HIAs), and to find the best strategies to reduce risks to human health.

The correlation between human health and sustainability/climate change that the report is based on is the fruit of more than twenty years of research activities in Europe. A group of researchers from the WHO European Centre for Environment and Health, using PubNet and ScienceDirect databases from January 2007 to September 2014, revealed how there is growing interest in the world of research on the direct and indirect effects of climate change on health and that a multidisciplinary approach and integrated assessments are increasingly preferred (Wolf et al. 2015). Since climate change is known to exist and that it consequently has demonstrated effects on health, the study in question maintains that it is time to apply rigorous policies by creating actions for effective mitigation and adaptation. Climate change is a threat to human health, but can also be considered a challenge to transition towards a more sustainable, healthier society.

This WHO study analyzed about 130 research projects in detail, observing some phenomena that are rather alarming for the European continent:

- mortality due to excessive heat is destined to increase, especially in Southern Europe, due to an increase in the frequency and seriousness of heat waves;
- the threat of extreme weather events could increase;
- the risk of fire, especially in countries such as Greece, Italy, Portugal, and Spain, could increase, with the relative threats to life, agriculture, and property;
- mortality caused by atmospheric pollution, as well as allergies, and illnesses due to the variation of pollen levels could increase, etc.

Despite the richness of the data and the studies analyzed, the researchers do not conceal the difficulty of this type of investigation because there could still be unknown associations between climate and health and there is still a lot to investigate and explore.

The knowledge of the effects of climate change on human health and therefore on the health of the city is today an integral part of national, European, and international reports and documents related to activating policies for mitigation (measures to reduce emissions) and strategies for adaptation (measures to reduce risks and adverse impacts).

Most actions that increase adaptation to the city climate and reduce the risks to health due to climate change refer to organizational models of the cities, technologies used for infrastructure and building safety, and the model of programming and planning social/production activities such as tourism and agriculture (ISPRA 2014).

Today local communities place greater focus on these themes and have a greater awareness that the risk is now—not in the future—so it is necessary to work to transform the risk into an opportunity.

But it is not only environmental sustainability that interacts with people's health and well-being. There is also social sustainability, which directly recalls the themes of economic sustainability and equity and inclusivity in European cities.

The "100 Resilient Cities" programme promoted by the Rockefeller Foundation, which supports cities in interventions aimed at urban resilience regarding environmental, economic, and social aspects, acts in this direction. This vision of resilience, as stated by Pietro Mezzi, goes beyond catastrophic events to encompass other stress factors such as high unemployment, inefficient public transport, endemic violence, and the chronic lack of food and water in some regions of the world (Mezzi 2016). All of this is promoted by providing expertise according to four well-defined paths: the institution of a Chief Resilience Officer, the willingness of experts to develop resilient strategies, the search for public, private, and service-sector partners, and finally, the exchange of experiences within the network of cities. Briefly, these actions provide support, direction, and exchange that is prefigured as a general planning strategy and monitoring in itineraries around which precise objectives can be pursued.

ICLEI—Local Governments for Sustainability is another platform within which it is possible to trace transverse themes pertaining to urban health and well-being, although always under the aegis of the term *resilience*. The same can be said for the C40 Cities Climate Leadership Group, which deals with six priority axes: adaptation and water, energy, finance and economic development, waste management, urban planning and development, and transport. In terms of sustainability and resilience, it should be noted how some of the objectives fall completely under the criteria of designing a safe, healthy city. They offer the possibility to work in an integrated, interactive way among the different policies and the specific definitions that the city assumes according to favoured—and in some cases, trendy—points of view: smart, healthy, sustainable, resilient, etc. Many of these, if not most, are undoubtedly similar common objectives that can be reached only by correctly managing urban planning policies and careful, wise direction in terms of planning (D'Onofrio and Trusiani 2017).

Current experiences show that there is still a lot to do, but it should be noted how there is not a one-size-fits-all plan when speaking about health and adaptation to climate change. Flexibility is required in order to understand what level of risk one is willing to accept.

In this sense, experiences within Europe (Barcelona, Copenhagen, Malmö, and Rotterdam, to name a few) and the rest of the world (Boston, Jakarta, Medellin, New Orleans, New York) constitute an interesting record for extrapolation with respect to some keywords running in the direction of health and quality of life in cities: environmental and social safety, public spaces and inclusive cities, and meaningful design references, both macro and micro.

By virtue of what is illustrated above, it seems appropriate to focus attention on some recent European experiences that highlight the interest of city administrations in reworking their territorial governance and planning/design documents according to criteria that favour environmental sustainability and resilience in a renewed relationship between health and urban planning. This is the situation in London, Malmö, and Turin. The choice falls on these cities because they represent an attempt to unite the terms *health*, *urban planning*, and *sustainability* on the different planning scales. The first two have drafted specific documents referring to urban health: London on the strategic level and Malmö on the strategic and operational levels. Turin, with its recent plan for the peripheries, descends to the neighbourhood level while being inserted in an urban vision. As well, all three explicitly declare and/or intersect the health/urban planning relationship by placing social sustainability at the centre of their operations as a fundamental prerequisite.

5.1 Box 1—London

In 2011, the city of London developed "Managing risks and increasing resilience. The Mayor's Climate change adaptation strategy". This is a strategic framework for adapting to climate change to improve the quality of life in the urban area and protect the environment. The document falls within a series of general strategies that define policies and actions based essentially on three founding elements: "Retrofitting London", "Greening London", and "Cleaner Air for London".

Synthetically, one can state that for "Retrofitting London", the modernization of existing buildings is fixed as a crucial element in addressing CO_2 emissions, reducing the use of energy and water, and offering new job positions and skills in addition to notable economic savings in terms of social costs.

For "Greening London", the starting point is the historical heritage passed down from the Victorian era, which can be used to draw inspiration to implement the greening strategy. This aims to improve the aspect and feeling of the city, making it more comfortable and attractive, with important effects for reducing noise and atmospheric pollution. "Greening London" also helps to render the city safer with respect to extreme climate events such as flooding, and contributes to the mental/physical well-being of people.

Regarding atmospheric pollution and "Cleaner Air for London", the actions adopted to improve air quality are focused and widespread: stricter regulations for low-emission zones, solutions and technologies that favour hybrid vehicles, hydrogen-powered buses, latest-generation green technologies, Source London (i.e., the first electricity network in the world to charge electric vehicles), the use of technologies to suppress dust and impede the circulation of PM10, the installation of green infrastructures to capture pollutants, and actions for awareness and monitoring on all levels.

Considering that London continues to grow and attract people and capital, these strategies are fundamental for a sustainable city that knows how to protect the environment and work continuously to reduce negative impacts, while improving the environment and quality of life. Following the 2012 Olympics, London has aimed for a sustainable low-carbon-based economy, convinced that this will also boost economic opportunities for the city in terms of jobs and investments.

Through these main cornerstones, the document introduces the theme of climate change and the need to work to manage risks, since the city is already vulnerable to extreme weather conditions such as flooding, drought, heat waves, and cold. It is necessary to reiterate the importance of working on resilience

not only in environmental terms but also regarding the economy, quality of life, and social equality. Many of the so-called vulnerabilities to climate impacts derive from urbanization. This makes it necessary to refresh green spaces and build common/community consciousness to increase resilience and improve the quality of life. Through this strategy, the administration provides a framework for collective action, implementing not only the administration itself, but also all Londoners, either associated or not, who can become active subjects in changing and/or facilitating the action and success of the strategies adopted.
The themes of risk are described point-by-point in the document, which considers the following questions: (1) Understanding the climate of the future: a summary of the projected changes to the climate that London will face; (2) Mapping adaptation: who is responsible for promoting and enabling adaptation and where are the critical gaps? (3–5) Understanding and managing the impacts. Covering the main impacts for London likely to result from projected climate changes (flooding, drought and overheating); (6–9) Analysing the impacts on crosscutting issues. Summarising the cross-cutting issues of health, London's environment, London's economy (business and finance) and infrastructure (transport, energy and waste); and (10) Implementing the strategy. Providing a 'roadmap to resilience', with a summary of the key actions, and an action plan.
Chapter 6 addresses the impact of climate change on health and highlights how the benefits and/or threats to health can be direct (heat waves, for example) or indirect (a hospital closed due to flooding). It also focuses on existing health inequalities, which will inevitably increase, in particular for vulnerable segments of the population. Finally, it examines relationships among health outcomes such as heat stress, cold stress, air-pollution-related morbidity and mortality, morbidity and mortality resulting from weather disasters, vector-borne diseases, water- and food-borne diseases, cataracts, skin cancer, and sunburn, as well as other known effects due to the weather/climate. Specifically, attention is focused on the following relationships:

- heat/cold stress and deaths due to heart- and lung-related diseases increase with hotter and colder temperatures, and heat-related illnesses (heat cramps, heat exhaustion, and heat stroke) and death increase during heat waves;
- air-pollution-related morbidity and mortality and weather are affected by air pollution concentrations; weather affects the distribution, seasonality, and production of air-transported allergens;
- morbidity and mortality resulting from weather disasters such as floods and windstorms cause direct effects (deaths and injuries), infectious diseases, long-term mental-health problems, and indirect effects (temporary limitations on access to health and social services);
- vector-borne diseases: higher temperatures shorten the development time of pathogens and increase the potential transmission to humans;

- water- and food-borne diseases caused by bacterial pathogens increase with increasing temperature. Increases in drought conditions may affect water availability and water quality due to extremely low flows;
- cataracts, skin cancer, and sunburn increase with more cloud-free days and higher temperatures and may encourage the potential risk of overexposure to UV radiation.

London's strategic document holds methodological and planning value of great interest because it focuses on the theme of risk, describes it in its structural components, and identifies general strategies and specific actions to implement over time together with public and private subjects. The roadmap to resilience is therefore composed of actions such as:

- promoting an integrated package of measures to enable and sustain long-term water efficiency
- integrating water efficiency into energy-efficiency retrofitting programmes
- promoting the capture and use of rainwater for non-food purposes to reduce the demand for water and reduce the risk of flooding
- improving the response to droughts
- prioritizing actions to target the worst-affected areas and most vulnerable communities
- managing temperatures by increasing green space in the city
- reducing the risk of overheating and the need for mechanical cooling
- ensuring that London has a robust heat-wave plan, health, economy, and infrastructure.

Each of these is specified in distinct measures, for example,

- managing temperatures by increasing green space in the city, i.e., developing 1000 ha of green space by 2012 to offset the urban heat island effect
- managing flood risk and providing biodiversity corridors through the city
- increasing green coverage in central London by 5% by 2030 and a further 5% by 2050
- managing temperatures in the hottest part of London
- increasing tree coverage across London by 5% (from 20 to 25%) by 2025
- enable the delivery of 100,000 m^2 of new green roofs by 2012.

This is also true of health, for example, where actions are aimed at ensuring that climate risks are addressed when commissioning and providing health and social services; in refurbishment programmes for the health and social care sectors as well, facilitating the provision of climate-risk information to borough Health and Well-Being Boards, and providing scalable examples of practical adaptation measures. This includes supporting a bid to the Technology Strategy Board for funding to retrofit a health building to improve its resilience to the impacts of extreme weather and climate change. Naturally, the specifics are valid for all actions, and each identifies the public and private actors involved in partnerships with the administration to reach the established objectives.

5.2 Box 2—Malmö

In 2012, the City of Malmö's Commission for a Socially Sustainable Malmö drafted the document "The city's spatial impact on health". The twofold objective of this report was to clarify factors in the physical environment that contribute to health inequalities and to highlight the opportunities offered by the urban-planning processes and their final results to obtain equable, favourable conditions to promote good health. The work is based on consolidated criteria of urban planning, convinced that it can influence and promote sustainable development and help to: reduce segregation among residential areas, improve trust, safety, and social opportunities, contribute to the correct location of schools in healthy, pleasing environments, contribute to a sustainable, mixed, and inclusive city, contribute to new economic and strategic structures, and use the experience and knowledge of people. In the document, some keywords exhibit particular relevance: *school*, *safety*, *health*, and *participation*.

The first part of the document underlines, almost as unavoidable assumptions, the importance of physical planning for social sustainability and the relationship between urban planning and health, delving into the concepts of the social environment, physical environment, and the importance of the physical environment to children and young people. Underlining the social and physical importance of the space in which one lives is a very evident sign that is launched in the areas of design, urban planning, and architecture as disciplines entrusted with transforming and managing the physical urban space. In fact, the physical environment is the background for all social interaction that takes place in a city. The design of the city is a basic condition for the health of its residents; the same social contacts, social control, and trust can be favoured or obstructed according to design criteria, for example, in a residential area. The role of urban planning is to influence internal social spheres via physical action. It is difficult to create a desired change in daily lifestyles on an individual level (e.g., increasing physical activity) unless the design of the neighbourhood encourages people to spend more time outdoors.

The document "The city's spatial impact on health" identifies some fields of action in which urban planning can intervene, and suggests some specific actions: (a) reduce segregation between residential areas by overcoming barriers, and creating good public transport and equal conditions; (b) improve trust, safety, and social opportunities by promoting meetings, social spheres, break normality, integration; (c) contribute to deliberately locating schools in good environments; (d) contribute to a sustainable, mixed, inclusive city through the mixed city, market forces, driving forces; (e) contribute to new economic and strategic structures through a new way of thinking; and (f) make use of people's experience and knowledge by using the planning process to promote increased local involvement through local initiatives.

The document constitutes the background for the Comprehensive Plan for Malmö, which was adopted by the Malmö City Council on 22 May 2014. This new comprehensive plan is a strategy carrying Malmö into the 2030s. The comprehensive plan is a representation of the City of Malmö's long-term vision for development and shows how planning can contribute to its implementation. The comprehensive plan consists of different parts: strategies, maps with planning guidelines, and an EIA.

The goal is to create a long-term, sustainable urban structure for a larger population, green growth, and continued development of the region's attractiveness. The ambition is to become a world leader in sustainable urban development, which implies a series of environmental challenges, among others. Achieving a socially balanced city in which everyone can enjoy good living conditions is at the basis of the plan's success. This is fundamental for the development of Malmö, even in relation to Copenhagen and Lund, with which it has formed an economically vibrant, attractive metropolis in the Øresund Region and which represents one of two incubators for competitiveness.

The plan aims to create a socially balanced city with good living conditions for all its citizens: a socially, ecologically, and economically sustainable city and an attractive place to live and work. The three areas of sustainability work in symbiosis and cooperation according to some well-defined priorities, such as: a close, dense, green mixed-function city; a regional driver of green growth and employment; the city as a venue for culture and democracy. The general objectives of the plan can be summarized as creating: (a) an appealing city that is socially, environmentally, and economically sustainable; (b) social balance and good living conditions; (c) economic dynamism and sustainability; and (d) a resource-efficient society and environmental robustness.

Between the lines of these strategies and general objectives, the local arrangement foresees a denser city within its consolidated limit, with a reduced consumption of land and resources, where public transport and bicycle use constitute the basic system of transport in the near future. Together with this, it recognizes the importance of urban spaces for the social development of the community and human contact, as well as stimulating democratic processes, inclusion/safety, and participation.

As mentioned above, the plan very much aims for social sustainability, that is, guaranteeing all the basic needs of each citizen, such as employment, education, safety, and health, but it also includes other aspects such as democratic participation, the sense of belonging, and opportunities for creative expression. A socially balanced city should also be an equitable city. Despite appearances, the plan explicitly speaks about a Malmö that is partially characterized by phenomena of segregation and social disparity: the differences in living standards and public health among the different districts of the city would seem to be notable. The task of the plan is to respond to this trend and free the potential offered by the city. The physical environment and the background in which social interaction occurs is therefore a basic condition for the quality of life in the city. Reiterating the concept present in the report that the physical, spatial, and formal organization of the city and its design influence the movements, habits, and life of its citizens, it is shown that urban planning, when working together with other disciplines/sectors, can affect the quality of the daily space and contribute to a city that is more socially cohesive.

If social sustainability plays an important, basic role, the economic sustainability of the plan is no less important. The plan foresees strong development of the city and its community, including businesses, with the aim of adding value in order to guarantee the region and population of Malmö sustenance and an income based on a strong post-industrial economy that can offer a series of work opportunities, redistributing the social balance and spreading well-being. Starting from the idea that cities and regions compete for people and capital, the attraction capacity of a city is influenced by a series of factors. These include the availability of adequate housing, functioning schools that are well-located with respect to residences and in close relation to favourable environmental conditions, attractive public spaces and cultural life, social cohesion and safety, and a high quality of assistance for children, who are fundamental in attracting families.

All of this corresponds to the organizational quality of a city and the role of urban planning is therefore also related to coordinating and suggesting correct means of managing public goods and collective economic resources. The socioeconomic perspective requires that the short-term objectives and priorities be calibrated with respect to the long-term process, especially regarding the responsible use of human and natural resources. In this respect, Malmö has set ambitious objectives related to the efficacy of resources and ecological sustainability. In collaboration with Copenhagen, Malmö intends to realize the first European cross-border carbon-neutral zone in the Øresund Region. The idea is to create a generator of green economic growth where growth and high quality of life go hand in hand. The economic dynamism of the Copenhagen-Malmö-Lund metropolis should also be of interest to the rest of Sweden and Denmark.

All of this is naturally interwoven with other sector policies, in this case related to infrastructure and society, forecasting the promotion of actions aimed at modifying movement habits on both the local and regional levels. All of this points in the direction of protecting the basic needs of future generations and conserving, in the long term, the productive capacity of the ecosystem as a basic requirement for human health.

Achieving the above-mentioned objectives requires the application of clear development priorities: a dense, green, and functionally mixed city that grows within the external ring road. In other words, the goal is a high-density city in the existing urban landscape that is capable of guaranteeing an efficient transport system that allows many people to abandon their cars in favour of transport with a low environmental impact, favouring at the same time the different design categories of slow mobility.

Another fundamental concept in constructing the plan is the concept of *proximity*: the Comprehensive Plan affirms that Malmö is a "close" city from many points of view: close to the continent, close to Copenhagen, close to the sea, and close to the country. This proximity is assumed as structural potential for further reinforcement, creating a flexible, multi-functional, and spacious city. The idea of proximity and "closeness" is applied to a series of priorities for the physical planning of Malmö: from community services that should be available in all parts of the city to public services such as schools, which should be situated close to residential zones. In the view of densification, residential areas are also close to private services and to details such as retail shops and other activities: the concept of proximity/closeness is united with the idea of a more compact city, which does not mean a less green city: the maintenance of green qualities in a denser city is fundamental and requires urban planning to suggest means and design solutions appropriate for managing free interstitial areas, existing parks, and designing new ones. Working on the concept that vegetation has properties of improving health, reducing pollution and noise in the city, and considering the view of densification, Malmö follows the path of "more vegetation of different types". The densification of vegetation in the central city corresponds to the dense city: this could mean the addition of new parks and/or the use of spaces, previously used for other things, for nature or for vegetation. Vegetation along streets should be implemented with the aim of notably increasing the number of trees along the streets and squares of Malmö. The strategy is to have a notable number of large and small parks, natural areas and urban squares that are strategically positioned, uniformly distributed, and connected by a series of green connections. One of the aims of the plan is that each residence have access to a park within one kilometer. Large interconnected parks and green areas become the cornerstone for recreational places, biodiversity, and ecosystem services.

The theme of green connections is tied to the strategies and objectives inherent in the system of travel and especially local transport. The plan clearly specifies that the solutions adopted should be centred on people because a well-functioning transport system can improve health, safety, and social cohesion. To this end, the objective is to create a more attractive, faster, more convenient, more comfortable, and safer transport system that is also more easily accessible to different social groups. The idea is an integrated public transport system with the other two priorities of walking and biking. In a denser city, with new buildings and connections, walking becomes even more important. Pedestrian traffic should be a priority, especially due to the contribution provided by pedestrians to urban life. Safety, comfort, maintenance, design, and easier orientation become important questions when encouraging people to walk more. The ambition of the City of Malmö to make bicycles the first choice for transport means implementing not only an existing network, but investing in new routes and working on the habits and behaviour of the population.

It is precisely along the public transport routes that the theme of the dense city makes way, which is intended as an existing city that should be integrated and densified particularly near railway stations and public transport routes. The different parts of the city should be designed to contain the widest possible variety of functions; functions such as residences, retail, social services, sport, culture, and offices can all exist in a single area. Concentrating buildings and activities both physically and functionally means containing land consumption in a highly urbanized municipality that still conserves elements of nature and natural resources. This means that it is even more important to protect existing natural environments such as fields, forests, and wetlands. Ecosystem services are fundamental for human existence and the capacity of a society to feed and maintain itself determines its future survival. An interaction between rural and urban areas is a prerequisite for a sustainable society. A denser, less diffuse city allows unique rural landscapes and natural environments to be preserved. Through conservation, development, and completion, rural and agricultural landscapes surrounding Malmö will become more attractive and increasingly accessible to city inhabitants.

The strategy oriented around the environmental landscape sets out three focus points: natural resources within the municipality should be used to promote long-term sustainability, border areas of the city with "elevated natural value" should increase, and the City should be restrictive about urban expansion into the surrounding rural and agricultural areas.

Naturally, the strategies outlined above act as the background for strategies to respond to climate change, sustainably manage waste and energy, and renovate buildings to save energy. In energy terms, Malmö has the declared objective of completely supporting the geographical area of the city with renewable energy by 2030, to better implement advantageous geographical conditions to produce renewable energy, heat and gas through wind, biomass, geothermal energy, and solar energy.

Concluding this discussion of Malmö's planning documents, it is useful to recall how all the strategies, objectives, and actions implemented reduce to a single objective enunciated in the Comprehensive Plan for Malmö, that is, public health is assumed for social development, well-being, and economic growth. This means creating opportunities for good living conditions and also encouraging healthy lifestyles. In this it is essential to encourage participation, and physical planning can promote it on the neighbourhood community level by actively encouraging the planning process with public representatives. In this view, the access and possibility of interaction in public spaces is a basic aspect of democratic societies. Malmö aims to create well-functioning public spaces of different dimensions and scales, local and urban worth, uniformly distributed throughout the city. Assuming the favoured point of view that a good place for meeting is accessible, welcoming to all, and preferably free, designing an accessible, welcoming city for all independent of sex, disability, sexual preference, or ethnicity is a very important aspect of urban planning. Beyond factors such as noise, pollution, and physical activity, health is strongly influenced by social factors such as participation, work, and societal trust. The plan repeatedly mentions that the planning and design of neighbourhoods can directly or indirectly influence these factors and contribute to improving the health and well-being of residents. In this sense, the plan recognizes a social role in urban planning and also challenges it to understand how to incorporate the above-mentioned factors—and more generally, the determinants of health—in the planning process, so they can be translated into criteria to design and modify the physical space.

5.3 Box 3—Turin

In the mid-1990 s, the City of Turin began to work on urban recovery and regeneration through a planning process that relied on policies, tools, and projects developed in the last fifteen/twenty years. This has radically changed the face of the city and its role in the European realm of urban competitiveness.
The post-Fordism era, with its six million (or more) square metres of decommissioned industrial areas (brownfields), has forced a reflection and a rethinking about the identity and future of the factory-city of the 1900s, as the previous assessor Ilda Curti has stated. It has also led to a rethinking of the large urban voids with transformation processes stimulated by being included—for 80% of the urban territory—under Objective 2 of the European Structural and Investment Funds. In addition, the winter Olympic games in 2006 provided a further impulse for urban transformation.

Together with correct planning, Turin has known how to utilize important public and private European and Italian financial resources. These have allowed the City to implement urban-transformation projects anchored in the idea of the urban plan as a tool to regulate and transform the city. Large transformations serve as the "backbone" in redistributing the use of decommissioned industrial areas and the new neighbourhoods created, i.e., the skeleton of the new Turin. Together with the crisis of post-Fordism, the same years saw an urban crisis that regarded the dense, inhabited, historical part of the city. "Quarters such as Porta Palazzo and San Salvario became emblems of conflict, which was often vindictive and hostile to new citizens, and citizens' requests for safety burst into the local political agenda" (Ilda Curti).

Also in this case, Turin addressed the problem via method and planning, initiating processes for urban recovery and regeneration, and mending the social fabric. This occurred in the semi-urban areas (Porta Palazzo, San Salvario) and in public housing neighbourhoods through new legislative devices available such as the programs for urban recovery (*Programmemi di recupero urbano*, PRU) the neighbourhood contracts (*Contratti di quartiere*), and actions for local development.

It is important to underline how, starting in 1997, Turin has dealt with the problem not only as a physical fact of urban regeneration, but especially as a factor to recreate social cohesion in territories that are unaffected by large urban transformations, addressing the theme of social safety in an apparently indirect way.

A large amount of regional, ministerial, and European resources have been employed in these areas to intervene on the outstanding nature of urban regeneration, understood over time as urban regeneration according to ordinary reasoning, which should be the culture of planning.

> The urban question requires complex thought. It means using a filter to read the city that allows it to be seen and act on in its structure and hardware. A common vice of those dealing with 'social' aspects is to consider them meaningless, but the hardware changes people's lives. At the same time it is necessary to invest in resources, energy, identity, and conflicts, that is, on the city's software. In a similar way, the vice of those dealing with infrastructure is to consider the software not as something important, but as an accessory. Only the combination of hardware and software allows the factory-city to be able to act, change, and hold itself together… 'Making a city' therefore means adopting strategic visions of development that have determined repercussions, which affect the lives of individuals living in the city. (Curti)[1]

[1] http://www.ildacurti.it/wordpress/chi-sono/.

In this scenario, Turin is working to understand various phenomena and act to govern them. It responds to the complexity of the city, opting for territorial projects that allow the urban fabric of the already-inhabited neighbourhoods to be regenerated. To do this, it implements the so-called "second-generation" policies about housing (that aim for living quality), policies to produce overall urban quality through services, *mixité*, and connections, and finally, policies in support of social proximity and in situations of social vulnerability.
Interest in Turin's experience lies in these latter policies. The social sustainability of urban transformations plays a determining role in providing tools to local communities and bringing cultural and economic resources together. The risk of gentrification in urban-transformation operations should be taken into account and governance should therefore be made with the aim of protecting those that live in the territory, providing tools to increase the quality of life, maintaining the fabric of proximity, and working on enhancing the identities, memories, and history of the territories. As Ilda Curti affirms, this implies mid- to long-term processes that require a global approach (looking at the city as a whole, improving the quality of life and promoting a transverse approach, integrating specialists and overcoming the compartmentalization of skill sectors, renewing systems and working styles), a territorial approach (connecting general policies and specific territorial areas, mobilizing social energy and resources and local institutions, enhancing the distinctiveness of the individual territories) and a design approach (participation, association, and partnerships are not built abstractly; it is in the development of projects that they are really found). In the last twenty years, the City of Turin has made ordinary what was once extraordinary in urban planning, focusing the process on the authority of public and private governance and subjects that have a collective responsibility for action in the long-term perspective.
In this view, within this scenario and as an integral part of strategies for local development and processes of governing the urban territory, the City of Turin approved the "*Azioni per le periferie torinesi*" [Actions for the Turin Peripheries] (AxTO) project in August 2016. The project relies on €18 million of state financing to initiate a profound transformation in the urban territory, with an overall budget of €41 million from public and private investments.
The AxTO project is organized around five thematic axes, which include 44 actions. The axes are: (1) public space, (2) home, (3) work and commerce, (4) school and culture, and (5) community and participation. The 44 actions established by the project are divided into five areas of intervention and refer to the following types of actions, established in the call by Art. 4, Section 3,

Letters a–e: quality of urban décor, maintenance and reuse of the public heritage, growth of territorial safety, enhancing urban services and welfare, and mobility and infrastructure upgrades. Statistical data related to three important dimensions of the city were used as the starting point to define the areas of intervention: the level of employment, the percentage of students in school, and housing degradation. Areas where at least one of the three indices returned a level of discomfort greater than the city average became part of the AxTO project. As explained in the project documentation,[2] the subject is an urban complex with a socioeconomic composition that is anything but uniform, within which there are also many neighbourhoods of cheap, public housing, and for this reason the city administration has opted for an intervention based on actions spread throughout the territory. These actions regard: the maintenance of houses, schools, infrastructure, green areas, and the land, the widespread support for establishing innovative micro-businesses, cultural production, and the social design of the urban community.

Of the five cited axes and the 44 actions expected, it seems evident that the AxTO project invests in the care and maintenance—not only physical—of the urban infrastructure, but it also invests in social capital, safety, and the capacity for resilience and urban sustainability, the strengthening of services and the creation of new models of metropolitan and urban welfare, promoted by public and private subjects.

The AxTO project proposes an integrated set of public and private investments for the 2017–2019 period, with the aim of opening the road to long-term regeneration and innovation in continuity with the most recent urban planning in Turin. The Turin Social Innovation programme and the related synergies with AxTO and other city initiatives (PON Metro, Co-City, Urban *Barriera*, Aurora Project) configure a set of projects capable of responding to emerging social needs and transforming innovative ideas into services, products, and solutions that are also capable of creating economic and social value to improve the quality of living and experiencing the city.

[2]General report "AxTO—Actions for the Turin Peripheries", developed by the city administration of Turin for participation in the "Bando per la presentazione di progetti per la predisposizione del Programma straordinario di intervento per la riqualificazione urbana e la sicurezza delle periferie delle città metropolitane e dei comuni capoluogo di provincia" [Call to present projects for the predisposition of the extraordinary intervention programme for urban renewal and the safety of the peripheries of metropolitan cities and Provincial capitals] from the President of the Council of Ministers of the Republic of Italy.

Inhabitants' participation, collaboration between citizens and institutions is a key element in the AxTO project. This is directly tied to some experiences already underway, such as the *Case del Quartiere* network[3] and the "Regolamento sulla collaborazione tra cittadini e amministrazione per la cura, la gestione condivisa e la rigenerazione dei beni comuni urbani" [Regulation on the collaboration between citizens and administrations for the care, shared management, and regeneration of common urban goods].[4]
Parco Dora and Incet are the two most important projects under AxTO. Located in areas of the city with the greatest evidence of three indicators of fragility and discomfort (unemployment, low level of education, housing degradation), they form the landmarks of AxTO. On the one hand, the completion of the Dora Park concludes the renovation of the broadest postindustrial area of the city; on the other hand, Incet (in the Barriera di Milano quarter, a fragile but vital "hemisphere" town[5]) holds the potential to transform and immerse social capital (it is not by chance that most of the co-financing by for- and non-profit subjects is concentrated here).
Through the AxTO project, Turin has formulated an inter-sector and integrated set of actions capable of addressing the main criticalities of the peripheral areas. From this point of view, in the last twenty years, the local system has solidified practices, knowledge, skills, habits, and attitudes towards cooperation that today constitute an important resource for urban policies. In the term *cooperation*, perhaps, lie the quality and innovative nature of the proposal based on actions that place humans and their relationships with the city at the centre.

[3]The *Case del Quartiere* are public buildings in which community functions, services, and opportunities for interaction and sociality are concentrated.

[4]Approved by the City Council in January 2016.

[5]These are parts of the city located in a ring around the centre. They are not central enough to attract investment, nor are they so degraded. These areas participate passively in change, while requiring micro-surgery interventions and new means of interpreting participation (Italiana 2007).

For example, the actions set out in Axis 1 (Public Space), are all characterized by a link with the processes of using and "appropriating" community places. These naturally go beyond attention to ecological-environmental aspects to secure the infrastructure, systems, and buildings, whereas in Axis 2 (Housing), the objective is to increase the safety standards of housing and the completion of interventions for renewal. The Incet community centre (Axis 3, Work and Innovation) is perhaps one of the most important actions for quality and innovation. The interventions are designed to be flexible, accessible, and aimed at supporting innovation to produce even short-term positive effects on the social and occupational plane of the territory through new rules of engagement between public and private entities based on the shared assumption of risk. In the actions under Axis 4 (Culture and School), the quality and elements of innovation are identified as thus: (a) in the willingness to overcome reasoning tied to simple "cultural events", to favour the spread of cultural events in the territory and occasions to socialize that have long-term solidification as an objective; (b) identifying participatory tools to identify initiatives to support and finance; (c) the willingness to promote the spread of opportunities throughout the territory, from schoolyards to mobile libraries. The actions established for Axis 5 (Community and Participation) inevitably touch the activation of what is established in the other axes in a transverse way: innovation and quality reside in experimentation with e-government and social mapping tools, in the shared management of common goods, in the city *Case del Quartiere* network to actions to contrast discriminatory behaviour and racism. The actions in Axis 5, permeating all the other axes, become a supporting structure of the entire AxTO operation and, as mentioned previously, can potentially constitute the innovative force of the programme, representing the social sustainability of the urban transformations.

The choice, then, to work according to axes lays out a strategic scenario that goes well beyond the programming flexibility of the experiences in the 1990s. This is because accurately defining the axes refers to a social process based on basic, fundamental societal questions such as school, education, work, culture, community, and participation which allows the political willingness to activate a process of growth and empowerment of the peripheries to be glimpsed in the view of improving the quality of life in the city.

To this end, the AxTO project, as stated in the project's General Report, promotes an improvement of urban quality through processes to care for and maintain public goods, the diffusion of services, and collaboration with citizens and economic operators. AxTO should be able to act as a catalyst for public and private resources, both economic and not. This appears in the articulated set of widespread investments in infrastructure, which can improve the layout of services and public spaces through a meaningful poly-centric rebalancing, from the creation of the Incet centre with the availability of new generative co-working spaces and the sharing economy, from supporting start-ups to projects for inclusion, cultural production, the associative fabric, and the service sector.

Social sustainability, lurking in the shadows for decades also in programming urban renewal to the benefit of environmental and economic quality, is the keystone of the entire AxTO and perhaps the most strategic of the three types of sustainability, potentially becoming a uniting element for a renewed relationship between health and urban planning. This is social sustainability intended as the capacity to guarantee conditions of human well-being (safety, health, instruction, democracy, participation, justice) equally distributed among citizens and increasingly viewed as a necessary prerequisite to realize economic and environmental sustainability.

References

D'Onofrio, R., & Trusiani, E. (2017). *Città, salute e benessere. Nuovi percorsi per l'Urbanistica.* Milano: Angeli.

Ispra. (2014). Focus su Le città e la Sfida dei Cambiamenti Climatici. Qualità dell'Ambiente Urbano. X Rapporto. Edizione.

Italiana, C. (2007). *La città abbandonata: Dove sono e come cambiano le periferie italiane.* Bologna: Il Mulino.

Kickbusch, I. (2010). Health in all policies: Where to from here. *Health Promotion International, 25*(3). https://doi.org/10.1093/heapro/daq055.

Mezzi, P. (2016). Le città per la resilienza urbana. In P. Mezzi & P. Pelizzaro (Eds.), *La città resiliente. Strategie e azioni di resilienza urbana e nel mondo.* Milano: Altraeconomia.

Pedersen, K. B., Land, B., & Kjærgård, B. (2015). Duality of health promotion and sustainable development—Perspectives on food waste reduction strategies. *The Journal of Transdisciplinary Environmental Studies, 14*(2). https://www.google.it/?gws_rd=ssl#q=Bransholm+Pedersen+et+al.,+2015. Accessed May 31, 2017.

Shiva, V. (2005). *Earth democracy: Justice, sustainability, and peace.* London: Zed Books Ltd.

WCED. (1987). *Our common future. World commission on environment and development.* Oxford: Oxford University Press.

WHO. (1993). Environmental health concerns of WHO. http://www.searo.who.int/indonesia/topics/ino-hen-2/en/. Accessed May 31, 2017.

WHO. (1999). *Environmental health indicators: Framework and methodologies.* http://www.who.int/ceh/publications/cehframework/en/. Accessed May 31, 2017.

WHO. (2016). *Health as the pulse of the new urban agenda.* United Nations Conference on Housing and Sustainable Urban Development, Quito 2016. http://www.who.int/phe/publications/urban-health/en/. Accessed February 14, 2018.

Wolf, T., Sanchez Martinez, G., Cheong, H., & Williams, E. (2015). The health effects of climate change in the WHO European Region. *Climate, 3,* 901–936. https://doi.org/10.3390/ijerph110606265.

Websites

London

For the experience of London, see:

https://www.london.gov.uk/WHAT-WE-DO/environment/environment-publications/managing-risks-and-increasing-resilience-our.

Malmö

For the document related to the Comprehensive Plan for Malmö, adopted by the Malmö City Council on May 22, 2014, see: www.Malmo.se/op.

For the document The City's spatial impact on health, Background report for the Commission for a Socially Sustainable Malmö, see:

http://Malmo.se/Kommun–politik/Socialt-hallbart-Malmo/Kommission-for-ett-socialt-hallbart-Malmo/Commission-for-a-Socially-Sustainable-Malmoe-in-English.html.

For the document referring to Malmö's path towards a sustainable future. Health, welfare and justice see:

http://Malmo.se/download/18.1d68919c1431f1e2a96c8e4/1491298331527/Malmo%CC%88kommisionen_rapport_engelsk_web.pdf.

Turin

For the experience of Turin, see:

http://www.comune.torino.it/rigenerazioneurbana/news/rigenerazione_urbana.htm.

http://www.comune.torino.it/rigenerazioneurbana/.

http://www.ildacurti.it.

http://www.comune.torino.it/benicomuni/news/axto-azioni-per-le-periferie-torinesi.shtml.

Chapter 6
The Crucial Point in Assessing Plans and Projects for Healthy Cities

Abstract Assessment of the impacts that the planning model and planning choices can have on the health of people and their quality of life is entrusted in many European countries and beyond to specific tools that are only rarely obligatory and integrated in ordinary planning tools. Their task is to evaluate the effects that the choices of plans, projects, and interventions can have on human health to correct them or orient their activation. In European countries, the HIA is required mostly for some categories of interventions, but rarely in the case of urban plans. Despite the evidence of certain risks, it can be of great assistance on the road to integrating urban planning and health, when the latter is carried out in conjunction with the former and with the urban project, later entering the phase of plan management through monitoring actions.

Keywords Health impact assessment (HIA) · Decision-support system · Sharing urban planning · Local democracy and participation · National best practices

Assessment of impacts that the planning model and planning choices can have on the health of people and their quality of life is entrusted in many European countries and beyond to specific tools that are only rarely obligatory and integrated in ordinary planning tools. Their task is to evaluate the effects that the choices of plans, projects, and interventions can have on human health to correct them or orient their activation. A publication by the WHO-Europe in 2014 entitled "Health in Impact Assessment" considers five types of impact assessment that hold this importance: the Environmental Impact Assessment (EIA), the Strategic Environmental Assessment (SEA), the Social Impact Assessment (SIA), the Sustainability Assessment, and the HIA. While the other types of evaluation also deal with the impacts on human health, the HIA is expressly dedicated to it.

In its current state, the HIA is not supported by binding legislation for European countries, as is the case instead for the EIA and the SEA. Although it was not compulsory, it was one of the central themes of Phase IV (2003–2008) of the European Healthy Cities Network, with the WHO's request for its habitual use to assess the impacts on health of projects, plans, and policies on a regional and local scale in order to provide recommendations on how to reduce risks, promoting the benefits to and monitoring the effects on health over time.

R. D'Onofrio and E. Trusiani, *Urban Planning for Healthy European Cities*, SpringerBriefs in Geography, https://doi.org/10.1007/978-3-319-71144-7_6

In European countries, the HIA is required mostly for some categories of interventions, but rarely in the case of urban plans. In addition, the HIA is used more on the local level than on the level of national plans and policies.

There are a series of reasons that have delayed the adoption of the HIA in urban planning.

In a recent article published in the Journal of Environmental Planning and Management, entitled "Environmental Health in the complex city: a coevolutionary approach", the authors discuss the complexity of the relationships between urban planning and health and identify three elements of weakness in the current evaluation procedure (Verbeek and Boelens 2016).

The first element regards the dynamics of the spatial transformations, for which impact assessments are often shown to be obsolete or incapable of addressing the continuous, accelerated evolution of urban phenomena. In a context of continuous change, territorial planning is beginning to experiment with design and regulatory solutions that are more adaptable and co-evolutionary (Boelens 2009; De Roo et al. 2012). Similar trends are slow in being realized in the area of impact assessments on health, where standards and regulations continue to be considered as something static and not, more appropriately, as components in a developing process that involves as many subjects as possible, among which solutions to problems are identified and negotiated.

The second element regards the difficult understanding of assessment processes. The EIA, SIA, and HIA are generally "black boxes" because they require specialized language that expands the gap between experts of different disciplines and between experts and citizens. Planners often do not have the technical skills necessary to understand the information provided by experts in other sectors, which makes it even more difficult to interact with citizens and associations, who often criticize the assessment procedure.

The third element regards the inadequate consideration for the social determinants of health. Impact evaluations essentially include environmental data that can affect personal health, but problems of health and well-being are generally not considered in a systematic way; they do not recognize the complex interrelationships between health, social conditions, and living spaces (Vancutsem et al. 2009).

Despite these effective difficulties, the HIA can be of great assistance on the road to integrating urban planning and health, when the latter is carried out in conjunction with the former and with the urban project, later entering the phase of plan management through monitoring actions. In the search for a new model of flexible, adaptable urban planning, the HIA could possibly indicate corrections and changes in path to make during the work. It would be able to correct the undesired effects of land transformations on the health of city inhabitants with the help of participatory paths and real sharing.

In the Foreword to this book, Patrizia Gabellini identifies some risks in applying the HIA. For one, it would introduce yet another assessment tool on top of already existing ones, further weighing down the process of developing plans and projects. In addition, its presumed inclination to control "everything" through sophisticated algorithms would mean using powerful databases that are often not available.

In our opinion, albeit with the limits highlighted above and without the pretence of relying on a valid tool on all occasions and in all contexts, the HIA has revealed a certain usefulness in the cases for which it was designed, i.e.:

- as a tool in support of decision-making (a decision-support system), whose goal is to achieve equality and equity in the health of all citizens;
- as a tool useful to promoting personal empowerment, in particular for the most vulnerable groups, favouring the involvement of private, public, and publicly controlled actors in search of an acceptable consensus (Sciences Po Toulouse 2015).

More specifically, its use would regard:

- the possibility of discussing different plan and design alternatives. Even where the HIA does not help to make decisions, it can still contribute to creating a debate that allows the possible impacts to be signaled;
- the possibility of developing forms of interaction and cooperation among the sectors of public administration and between these and citizens and interest holders;
- the possibility of increasing local democracy and participation. The HIA could be configured as a transparent tool to gather and show the effects of the different alternatives of a project, providing the possibility to share the choices with the population.

There are many important HIA experiences in Europe. We have selected two regarding its application: the City of Rennes and the restoration of the Pontchaillou railway station, and the city of Bristol in the case of Greville Smyth Park (see Appendix 2).

In the case of Rennes, the HIA represented the first experience in France in the area of an urban regeneration project. This initial application caused some critical and some potential aspects of this tool to emerge. Its application confirmed the general lines of the project to restore the Pontchaillou station and the surrounding neighbourhood, but it also contributed to providing a concrete list of new possible solutions and to reinforce the probability that the project could have a positive impact in terms of personal health (Tollec et al. 2013). The HIA also proposed the objective of allying the different players (experts, communities, politicians) around the central question of health and the quality of life, developing a method centred on a multi-criteria network. This network allowed the different determinants of health to be investigated in relation to the design proposal in order to best evaluate the choices of the politicians responsible and to provide useful recommendations to promote the success of the project. The inter-sector approach worked, while requiring a large effort in coordination and cooperation among public and private players, which is not always easy to achieve, especially when different actors use different languages. The language difficulty was overcome because preliminary meetings were held in which all interested subjects were taught the common definitions, such as the concepts of *health*, *air quality*, and *noise pollution*.

What did not work was the involvement of citizens and users (people interested in the project, users of the TER Bretagne railway network, users of the Rennes University Hospital, local residents, and students). A further difficulty, this time highlighted by the stakeholders, regarded the access to information, which was often difficult to collect. This led to problems such as: the increase in time to draft the HIA and compromising the quality of the process and the results.

In the case of Bristol, the success of the HIA, in contrast, regarded precisely the involvement of the population in the project. The experimentation dealt with Greville Smyth Park, a green area situated to the southeast of the city centre, which was frequently subject to vandalism by youths. The city administration, in collaboration with the non-profit organization Friends of Greville Smyth Park (FroGS)[1], defined, in the Greville Smyth Park Improvement Plan, design choices aimed at converting the area's problems into strong points, encouraging youths to use the park in a positive manner. Recognizing youths as key users, the organization established a participatory process aimed directly at them. There were many initiatives implemented (events and demonstrations related to art, etc.), and many players contributed economically to realizing the new park and the participatory process.

References

Boelens, L. (2009). *The urban connection: An actor-relational approach to urban planning*. Rotterdam: 010-Publishers.

De Roo, G., Hillier, J., & Van Wezemael, J. (Eds.). (2012). *Complexity & planning: Systems, assemblages and models*. Farnham: Ashgate.

EHESP/DGS, Roué-Le Gall, J., Potelon, J. L., & Cuzin, Y. (2014). *Agir pour un urbanisme favorable à la santé, concepts & outils*. www.ehesp.fr/wp-content/uploads/…/guide-agir-urbanisme-sante-2014-v2-opt.pdf. Accessed June 6, 2017.

Fehr, R., Viliani, F., Nowacki, J., & Martuzzi, M. (2014). *Health in impact assessments*. WHO Regional Office for Europe. social-sante.gouv.fr/IMG/pdf/02health-in-impact-assessments-final-version.pdf. Accessed June 6, 2017.

Sciences Po Toulouse, Iferiss. (2015). *L'evaluation d'Impact sur la Sante (EIS). Analyse compare de dé marches territoriales étrangères pour améliorer la santé e l'équité en santé.* http://www.iferiss.org/images/IFERISS/2015_rapport_eis.pdf. Accessed June 6, 2017.

Tollec, L., Roué le Gall, A., Jourdren, A., Auffray, F., Jabot, F., Vidy, A., Thébault, P., Mordelet, P., Potelon, J. L., & Simos J. (2013). L'Évaluation d'Impacts sur la Santé (EIS): une démarche d'intégration des champs santé-environnement dans la voie du développement durable. Application à un projet d'aménagement urbain: la halte ferroviaire de Pontchaillou à Rennes. *Developpement durable & territoires* (Vol. 4, no. 2), Santé et environnement. https://developpementdurable.revues.org/9815. Accessed June 6, 2017.

Vancutsem, D., Gee, D., Gossop, C., Hoyer, U., Jarosinska, D., Laconte, P., Schrenk, M., Hemis, H, & Seidl, R. (2009). In J. Colman (Ed.), *Urban planning and human health in the European city*. Report to the World Health Organisation. International Society of City and Regional Planners

[1]The organization aims to improve the environment and structures of Greville Smyth Park and the surrounding areas, through actions related to safety and accessibility. In addition, it looks to initiate actions to encourage local residents to care for the park and to act responsibly.

(ISOCARP). https://isocarp.org/app/uploads/2014/08/WHO_report_final_version.pdf. Accessed June 6, 2017.

Verbeek, T., & Boelens, L. (2016). Environmental health in the complex city: A coevolutionary approach Environmental health in the complex city: A co-evolutionary approach. *Journal of Environmental Planning and Management*. https://doi.org/10.1080/09640568.2015.1127800.

Chapter 7
Comparing European Cities on the Road to Integrating Health and Urban Planning

Abstract The effective difficulty of intersecting the themes of health and well-being within urban-planning tools and how much, instead, this relationship is desirable even for the goals of promoting sustainable development, suggests investigating the theme with the aid of some current experiments in European cities, selecting some key questions around which the wager on integration is made. The comparison regards essentially medium or small cities (Belfast, Bristol, Ljubljana, Odense, Pécs, Poznan, Rennes, Turku, Modena, Udine); an inter-municipal association (Provence-Alpes-Cote d'Azur, PACA); and two metropolitan areas (Turin and Bologna). The results of this research mainly regard the content of the urban plans from the strategic to the operational levels with reference to the themes of health and well-being. In addition, aspects connecting the plans' choices to the realization of interventions to design the places and empower local communities are investigated. Appendix 2 presents three of these experiences: Bristol's Parks and Green Space Strategy; Rennes' Restructuration de la Halte Ferroviaire de Pontchaillou; Healthy Poznan—Development Strategy for the River Warta; and the Hirvensalo District Master Plan.

Keywords Strategic planning · Local masterplan · Inspiring practices · Urban health · Well-being

The effective difficulty of intersecting the themes of health and well-being within urban-planning tools and how much, instead, this relationship is desirable even for the goals of promoting sustainable development, as emerged in Chap. 5, suggests investigating the theme with the aid of some current experiments in European cities, selecting some key questions around which the wager on integration is made.

While aware that it is not possible to generalize, because each city is a separate case, the scope of the comparison is to:

- focus attention on the basic questions regarding the possibility of managing the themes of health and well-being within local urban plans;
- verify the impediments that are interposed and if good practice can provide useful suggestions to overcome the objective difficulties encountered.

R. D'Onofrio and E. Trusiani, *Urban Planning for Healthy European Cities*, SpringerBriefs in Geography, https://doi.org/10.1007/978-3-319-71144-7_7

To do this, with the aid of a group of young researchers[1], a comparison between some European cities was made.

This choice purposely regarded different cities pertaining to rather different countries, traditions, and regulatory regimes regarding urban planning and public health. Specifically, the following characteristics were chosen: cities that for some time have addressed the themes of health and well-being in the presence of national standards and regulations that study and try to favour this integration; cities that are developing integration experiences, even where there are not general consolidated procedures, but only directions; and finally, other cities that are new to the European Union, while being interested in following the path of integration, show evident cultural gaps and administrative difficulties.

This comparison regarded essentially medium or small cities (Belfast, Bristol, Ljubljana, Odense, Pécs, Poznan, Rennes, Turku, Modena, Udine), an inter-municipal association (Provence-Alpes-Cote d'Azur, PACA); and two metropolitan areas (Turin and Bologna). In this section, only the main results of this research will be presented. The results mainly regard the content of the urban plans from the strategic to the operational levels with reference to the themes of health and well-being.

In addition, aspects connecting the plans' choices to the realization of interventions to design the places and empower local communities were investigated. Appendix 2 presents three of these experiences:

- Bristol's parks and green space strategy;
- Healthy Poznan;
- The Hirvensalo District Master Plan.

7.1 The Role of the Plans

Regarding the comparison on which the first part of the research was focused, the main keys to reading regarded:

(1) the capacity of the plans to incorporate the themes of health and well-being, with reference to the following areas of interest:

   - synergy and consequentiality among strategic, local, and detailed plans in taking charge of the themes of health and well-being;
   - aspects addressed (determinants of environmental and social health);
   - capacity to involve the different sectors of the public administration (with particular reference to the sector of public health) to formulate common plans and projects.

(2) the means of involving the local community and interest holders in formulating plans, projects, and the HIA, with regard to:

[1]Ph.D.: C. Camaioni, P. Pellegrino, M. Tolli; Ph.D. student: F. Stimilli.

- forms of participation;
- the effective capacity to orient processes, influence decisions and the monitoring phase, and control the results of the transformations.

(3) the effectiveness of the solutions implemented, with reference to:

- the existence of recurring "impediments" that threaten the possibility of effective integration of the themes of health and well-being;
- the effective possibility of solutions to the problems, to be applied in other contexts and experiences.

For Bristol, Belfast, Odense, Turku, Pécs, and Ljubljana, the experiences studied in the research referred to the set of plans and projects developed by the municipalities in which attention for the themes of health and well-being were widely made explicit. In the case of Rennes and the PACA Region, reference is made to the HIA procedures activated for the Pontchaillou railway station in the case of Rennes, and for the Integrated Urban Projects (*Projets Urbains Intégrés*, PUI), in the case of the PACA Region.

In Italy, the direct/indirect capacity of urban plans to encompass themes of health and well-being is generally seen. Turin and Bologna are particularly evident among metropolitan areas, while Udine and Modena and the medium-small cities generally display interesting point-like experiences that intersect planning and sometimes direct involvement.

By comparing the different experiences, some critical points and some innovations emerged for urban planning. In reference to the first interpretational key, substantial differences among the different case studies were revealed. The experiences of Belfast, Bristol, Odense, and Turku show interesting steps down the path to integration. All levels of planning are involved (from strategic plans to local plans to detailed plans); from general strategies the step is made towards actions and interventions according to a logical, strongly interconnected sequence.

This is what occurs, for example, in the case of Belfast. The long-term vision of development for the city, established in the Local Development Plan (LDP) proposes a healthy lifestyle based on physical and emotional well-being and on the reduction of health inequalities. This vision is better specified in the Belfast Agenda 2017–2021, which promotes concrete actions to improve the quality of life in neighbourhoods to reduce social inequalities and to promote healthy lifestyles for all ages. More operational documents follow, such as the City Centre Regeneration and Investment Strategy, which delineates a programme up to 2030 that connects strategies, actions, and related investments for the urban regeneration of the central and surrounding areas of the city.

In the case of Odense, it is in the strategic plan (*Planstrategi 2015*) that a series of priorities for the city are selected, in close relation with the theme of health. It is, however, the local plan (*Kommuneplan 2016–2028*) that establishes the guidelines and specific objectives to make the city denser, develop the system of green areas (urban parks, neighbourhood parks, tree-lined avenues, gardens, green façades and roofs), and intervene on traffic.

In the case of Ljubljana and Pécs, the integration was rather difficult. This is evident in the recourse made to sectoral projects that are not completely integrated with each other or with urban planning. The local plan reveals an insufficient attitude to address programmes and projects in other sectors of the public administration, especially the health sector.

In the case of Ljubljana, for example, the local strategic planning document (*Trajnostna urbana strategija mestne občine Ljubljana 2014–2020*) makes both broad and specific reference to integration with the themes of health and safety by identifying development objectives centred on the promotion of a healthy lifestyle, the provision of sports centres and recreational areas, the safety of weak and socially disadvantaged segments of society, the need for inclusive urban development, etc. In practice this is entrusted to a collection of sample interventions that cannot easily be contextualized in a single integrated project.

In the case of the French cities, integration is pursued through the HIA. It is used as a tool to evaluate the detailed plans in the case of the PUIs in the PACA Region and the urban development projects (*Projets d'aménagement urbain*) in the case of Rennes.

For the Italian cities, in terms of synergy and consequentiality among the strategic, local, and detailed plans, in taking charge of the themes of health and well-being, it is clarified that the themes of health almost always enter urban plans indirectly and transversely. This occurs through stated general policies, strategic objectives, and specific projects that almost always refer to urban regeneration or through experimental or promotional point-like actions activated by the Healthy Cities Network. Among the metropolitan areas, the experiences of Turin and Bologna are important. Here, this capacity can be traced to the consolidated daily management of the city based on integration and participation as well as to some practical innovations and prudent projects in the medium and long terms. In both cases, naturally, the plans and projects implemented on the macro and micro levels in the specific field of urban planning and design refer to the themes of resilience, environmental sustainability, and the smart city, just to name a few. In the field of urban planning, the now-consolidated synergy is with environmental questions, with participatory practices and questions tied to mobility, while synergy with the determinants of health is still struggling to provide significant input for the discipline, referring questions to the prerogative of the social/health sector and welfare. Signals of change can be seen in some experiences within the Healthy Cities Network, as with the Public Health Agency (*Agenzia di Tutela della Salute*, ATS) in Bergamo, where local input and experimentation appear on the neighbourhood level. There are a few exceptions to what is described above, but it is worth noting the Association of Bazzano Municipalities (*Associazione dei Comuni dell'Area Bazzanese*) in the Emilia Romagna Region, where the dialogue grows out of an intense collaboration, first on the regional level and then on the local level, between the health and the territorial planning sectors, which is concretized by experimenting with the regional guidelines on a local plane (Associated City Structural Plan, *PSC Associato*).

With regard to the aspects considered to improve health and the quality of life, there is a certain uniformity among the different experiences. The determinants of health investigated the most regard lifestyles and environmental factors, with particular reference to: green areas, sustainable mobility, safety, and the quality of life of vulnerable populations, city densification, urban regeneration of the most degraded areas, accessibility, the quality and functionality of housing, air quality improvement, extended pedestrian zones and restricted traffic zones, the creation of new walking/biking paths, etc.

In addressing these themes, interaction with other sectors of public administration, especially the health sector, does not always occur with the same incisiveness. In the case of Bristol, for example, the interaction was initiated especially in the phase to build the initial tools, making use of a protocol in which the terms of the partnership were defined. The latter foresees actions in the field of housing (acting on architectural quality, the liveability of spaces, accessibility, type, and density); the quality of public and green space; and accessibility for all. In the case of Turku, strategic agreements were made. Beyond involving local actors (communities, associations, business representatives, etc.), these agreements are real working tables for cooperation among the different departments in the municipal structure, residents, entrepreneurs, doctors, and cultural and sports associations. In the case of Rennes, the dialogue among administrative entities, planning tools in the health sector and urban planning themes are evidenced through the local health observatory, the local health plan, and the local health contract. Among the objectives of the Local Health Plan, the closest to urban planning, is "Promouvoir un Cadre Urbain Favorable à la Santé" [Promoting an Urban Framework Favourable to Health]. This plan expressly singles out the general urban plan and the detailed plan to improve the quality of life and walking/biking to develop practices that respect the environment and health to guarantee environmental quality and prevent pollution of the land.

In the cases in Italy, the theme of quality of life can generally be traced to recently developed urban planning instruments, even if in practice the interventions and actions are struggling to be applied effectively. The theme quality of life is found in almost all projects to regenerate peripheral areas, not least in those approved for financing by the Italian government programme launched in 2016 which crosses health, social, environmental, and building security, for example. In the case of Bologna, instead, the quality of life already appeared dominantly in the general urban planning tool some years ago. The quality of life permeates the structural idea of the seven cities of Bologna and projects in the operational plan on the micro level. For Turin, it becomes a present and future theme in actions for the peripheries of Turin (AxTO), as described previously. At the same time, it can be considered a mid- to long-term wire running through the Piedmont regional administration in the process/step from industrial city to cultural city. On the legislative level, mention is made in the introduction to the HIA and some regional experiences, but in fact the recent approval does not allow an exhaustive critical framework of assessment of the results to be developed. What is certain is that the HIA appears as a further weight in the planning and assessment system, and this could block not only its application, but also the normal process to develop and approve urban plans. Recent suggestions to involve the content of the HIA in the already consolidated—and required—SEA seem to confirm this.

The second key to reading regarded the means of involving local communities and interest holders in formulating plans and projects that regard health and well-being. The usefulness and effectiveness of this involvement is shown clearly in the habits and procedures used and experimented with by the administrations to follow the path of participatory democracy and citizen empowerment.

In the case of Bristol, Belfast, Odense, and Turku, the involvement of the population and stakeholders occurs through different means: from promoting popular and cultural events, to opening co-planning tables, to the realization of different forms of permanent consultation and real paths in community planning. In addition, it is often the citizens themselves that form grass-roots community groups that are involved in sharing information, skills, know-how, and experiences. In Bristol, for example, a network of neighbourhood volunteers was created: the Neighbourhood Planning Network (NPN). This is dedicated to increasing the trust and effectiveness of community groups in carrying projects forward and in actively taking part in the planning and design of urban spaces. In addition, the city, in constructing the Neighbourhood Development Plans, directly involved citizens by organizing coordinated planning tables to outline policies and actions in response to local needs and priorities. In these tables, health and the quality of life play a central role.

In the case of Belfast, the permanent involvement of the local population occurs with the support of a regulation—the Statement of Community Involvement—and is based on community planning paths. To realize the specific themes of the WHO Belfast Healthy City, the Regeneration and Healthy Urban Environment Group was created. It promotes programmes for urban regeneration with the collaboration of partnerships composed of organizations and/or local entities that deal with overcoming social inequalities and, more generally, activating integrated models of planning with the theme of health.

In the case of Odense, the promotion of health specified in the objectives of the city health policies is based on many opportunities for cooperation that are created among the City, local communities, educational institutions, health operators, etc. This occurs effectively through the institution of community centres, which are a mixed sort of health, social, and cultural centre.

In the case of Turku, the population's involvement in building the strategic plan and impact assessment regarded the evaluation of a vast range of opportunities (scenarios) for each district of the city. Cooperation in constructing these scenarios works to increase the sense of the local community, reinforcing social awareness, preventing problems, and increasing the perception of safety in the city.

In the case of Ljubljana and Pécs, local involvement is more fragmented. This occurred mostly in the initial part of the process to then be interrupted. In Ljubljana, for example, participation occurred before the objectives and priorities of the projects were defined, and often regarded not all social classes, but only the upper-middle class.

In the case of the PACA Region, involvement of the local population and stakeholders accompanied the HIA path and was shown to be very useful in becoming familiar with and taking stock of the different particularities of the territories. This saw the activation of workshops, individual interviews, and focus groups. In the case of Rennes, citizen participation is an important component in the HIA process, but is also one of the main problems. The means of involvement, in fact, are not regulated by guidelines or minimum criteria, and this has led to insufficient involvement of the community (only one meeting), which was not ready to become fully involved.

In the case of the Italian cities, it is interesting to note a change—at least in this initial phase of experimentation—in the means of acting and in proposing the participatory process. For some years and in the wake of actions in the City of Bologna, some cities have adopted a *Regolamento dei Commons* [Commons Regulation] that substantially changes the administration/citizen relationship. There is no longer "post-it" participation, as the assessor of Bologna confirms, but responsible participation in a process of co-design and co-monitoring of the results of the transformations. Experiences on the neighbourhood level reveal an interesting social framework of bottom-up movements and renewed interest in the city as a common good, in addition to showing the "neighbourhood" as a field of investigation and planning unity for experimentation in the field. New participatory forms take shape as collaborative pacts. In Turin, for example, we recall the Health Centre (*Polo della Salute*). This is a space where an overall vision can be generated and where actions to aggregate the different entities involved in health planning can be reinforced. The portal http://www.comune.torino.it/pass/salute becomes an agora dedicated to urban health and is configured as a physical and virtual place for meeting, debating, organizing events, raising awareness, education, training on the theme of well-being, etc. for all those operating in the different city territories on themes involving well-being and the promotion of health. Faced with a diffuse, renewed, and desired participation, the forms activated by Bologna and Turin are highlighted because they are directly and/or indirectly posed as structures to promote, train, and educate about the theme of health, and serve as references for the daily management of creating the city through community forms (the neighbourhood and beyond), that fully enter the planning and transformation process for the urban environment.

By applying the third key to reading, it was shown that one of the reasons for success of some of the experiences examined regards the specification of clear strategies in terms of health and well-being, which become a constant reference for the different actions and projects that administrations are capable of implementing over time.

In the case of Bristol, for example, the Bristol 2020 strategy, which projects that by 2020 the city will become one of the top 20 cities for sustainability, is directly and indirectly tied to themes of health, with the aim of characterizing all the urban policies. In the same way, this occurred in the city of Pécs with the Borderless City project in 2010, which triggered a further series of interventions to improve the urban environment and the quality of life of citizens.

One of the most controversial themes in these experiences is the application of the HIA.

This was verified above all in countries that have been addressing the themes of health and well-being within urban policies for the least amount of time. In the case of Pécs, for example, this is due to the difficulty of the administrative machine and local technicians to address the assessment of plans and projects. Despite the great effort made on the level of training and publishing the HIA, in a collaboration between the city administration and the Healthy Cities Foundation, the HIA has not achieved the desired results.

Another problem emerged in the case of Rennes. The inter-sector approach, which is one of the most important aspects for the validity of the HIA, required great effort to coordinate and create cooperation among the actors, and this was often very complicated to pursue. In the case of the PACA Region, difficulty was instead found in accessing information. This was due to the fact that there were too many communities involved and it was necessary to select only some actions for application of the HIA. A further difficulty regarded the assessment of impacts, which were not always easy to interpret in the case of "neutral" impacts.

In the case of Turku, while good integration was seen between urban planning and the themes of health and well-being, with an important role played by the assessment tools in determining the plan choices, the risk that became clear was that along the path, the matrix of these tools, which was aimed especially at assessing the impacts on health and well-being of the urban populations, was forgotten, and that the choices were also affected by other events that were difficult to control.

In the cases in Italy, there is intense activity to promote, raise awareness, and experiment with point-like actions that lie directly outside the urban planning and health relationship, and therefore outside urban plans, which constitute in some cases a preliminary experimentation to then be included as planned actions in the tools in progress. In some cases, this may refer to projects promoted and realized in the area of Eurodesign, as in the already-cited Heat application, or in local experiences on the theme of sustainable mobility and green areas in the case of the Pedibus in Udine. Health, urban planning, and economic aspects come into play in the project and represent an interesting experiment to consider. For the rest, the effectiveness seems to reside in long-term actions for virtuous communities, but they are almost never traceable to daily practice or extendable to wider territories. The Emilia Romagna Region, and many of its municipalities, perhaps constitutes an exception in this sense.

The cases of Bologna and Turin represent meaningful points of reference on the different scales of urban planning, but it should also be underlined in these cases that a united health/urban planning path is lacking. The difficulty resides in integrating the different sectors, the difficulty of dialogue, and perhaps the absence of a common vocabulary on which to work, as well as in the insufficient economic resources to invest in new professional figures that know how to ease communication between experts in different sectors.

From the comparison made, some considerations have been extrapolated that may be useful in order to widen the debate on the integration between urban planning and health, including:

- the effectiveness of a clear system of objectives and strategies expressed in terms of health and well-being in living environments. This is useful in favouring the integration of plans on different levels, plans and projects to feed the interaction among the various sectors of public administration/public health agencies and urban planning, and among the plans and programmes they produce;

- accompanying the regulatory plan of a system of sectoral tools and policies capable of responding to the different reading of needs in terms of health and well-being;
- the importance of involving the local community and stakeholders through bottom-up processes. The most effective processes are those in which citizen organizations work side by side with public institutions rather than processes guided from above, as it is more useful to address select groups of representatives rather than halls that are too crowded. Actions for concrete involvement are to be preferred, which give direct responsibility to the citizens (a relationship based on "reciprocal learning among requests and different interests is desired) and which also extend to the phase to monitor the plans and projects;
- the need for a common language among experts in different sectors, to be found first within urban-planning tools to ease disciplinary misunderstanding;
- the awareness of the community, stakeholders, professionals, and politicians of the close relationship between plans, projects, determinants of health, and quality of life in the city. This relationship is not often seen, and partial responses that are mostly ineffective continue to be provided when confronting complex problems.

7.2 The Landscape, from Plans to Projects to Interventions

An additional investigation made regarded the step from urban plans to projects to improve health and well-being, to the realization of interventions and their management. Particular reference was made to:

- the methods and operational tools referred to in the plans or that the city provides to orient the design of places;
- the procedures implemented by the plans or to which the plans refer to make local communities responsible for caring for the living spaces.

The following case studies were considered: Bristol's Parks and Green Space Strategy, the Development Strategy for the River Warta in Poznan, the Hirvensalo District Master Plan in Turku. Appendix 2 is dedicated to these projects.

In the case of Bristol, the Parks and Green Space Strategy developed a clear methodological approach to build knowledge and design interventions, to implement the "Bristol Parks" 2008 strategy, drawn and expanded from the local plan. This methodological approach, as described in Appendix 2 (prepared by Piera Pellegrino), established:

- an extremely in-depth familiarizing and assessment phase based on consistency and the level of quality of green areas and their state of maintenance. The main objective is to define an assessment model to estimate the costs of the interventions;
- a consultation phase composed of different tools and participatory phases that also included the EIA. The consultation phase served to identify the objectives to pursue and to identify the devices to consider.

In addition, to guide the operational phase of the interventions, the following was set out:

- recourse to qualitative and quantitative standards for the green areas and measurements to improve accessibility to these areas;
- the active involvement and responsibility of local communities as the main creators in improving the areas (Greville Smyth Park).

In the case of Poznan (Appendix 2, prepared by Flavio Stimilli), the Development Strategy for the River Warta favoured the activation of some strongly innovative initiatives to encourage the local population to participate in projects and concrete realizations, including the Green Backyards Initiative, which is proposed to make backyards of tenement houses more attractive. The programme (formally "Change your Backyard", *Odmień swoje podwórko*), involved inhabitants in designing and creating small green spaces in their backyards. Soil, plants, and expert advice were provided by the city. Apart from greening backyards, the project also aims to improve social integration and encourage inhabitants to work for the common good. The stakeholders involved include local citizen groups, the NGO TASAK, the local District Council, and homeowner associations.

Citizen involvement occurred by applying a cutting-edge methodology. In 2014, the Public Participation Geographical Information System (PPGIS) was used to create an interactive portal based on GIS technology, through which the City could communicate and collaborate with residents (appropriately trained and informed), to reach a shared draft of the local development plan for the area around the Kasprowicz Park.

In the case of Turku, the HIA was used in the Hirvensalo District Master Plan as a tool to identify the best model to update the local plan. Based on three structural models presented to the local community within some workshops, relative value judgements were attributed to the three different design proposals. Each model was assessed with reference to the pros and cons regarding quality of life, housing, employment, commercial services, public services, traffic, noise and emissions, climate, technical services, the environment and the landscape, the cultural environment, etc. All three models were shown to have positive and negative impacts on health and well-being. The proposal chosen to revise the Master Plan was a synthesis between the first and second models: traffic was organized as set out in the first model, while the settlement structure relied on indications in the second model.

In this case, as was illustrated in Chap. 6, the HIA played a fundamental role in building a decision support system (DSS) that led to the identification of optimal design solutions.

What emerges from these four case studies is the exploratory and mediative function of the project aimed at improving well-being and health in cities, stimulating and activating the social fabric and enhancing the resources that the community—the bearer of "needs"—can implement.

The participatory project of the spaces for health and well-being is capable of transforming citizens from passive users into active users of the city, empowering them to address living spaces and creating a model to control and monitor the spaces that can be replicated in other contexts. The management of this type of project aims to raise awareness, reformulate the problems, open the debate to the largest possible portion of the population—organized or not—animate the relationships between social groups, changing the normal balance of administrative powers. In this way, the project produces new levels of awareness and responsibility in the name of health and well-being, which can create conflicts, but which, when analyzed and systematized, can only be positive and useful for a greater use of living spaces.

In the last Part of this book we reflect on some case studies in Europe and other parts of the world in order to explore the possible desired effects of this liaison.

Part III
Planning and Designing Healthy Cities and Communities

Chapter 8
Improving Health Through Community Urban Planning

Abstract On the one hand, contemporary cities are called to come to terms with the local community's renewed role as co-manager, co-designer, and co-producer of the living spaces. On the other hand, they must consider the role that urban design can play in designing health-based cities. In order to highlight and reflect on these themes, this chapter draws on some best practices: New York and Toronto for North America and Medellín, Santiago de Chile, and Porto Alegre for South America. These themes hold notable importance in urban policies and, albeit from different points of view, aim to improve the quality of life in these cities. The experiences referred to in this Chapter show how urban planning favorable to health amplifies the need to draw on local and experiential knowledge of the urban environment reflected in the community.

Keywords Urban design · Community urban planning · Health-based cities Resilient communities · Quality of life

As anticipated at the end of Part II, this book concludes by focusing attention on the virtuous relationships that can be activated between the local community and urban designers in creating healthier, more equitable cities.

Cities with such aspirations are called, on the one hand, to come to terms with the local community's renewed role as co-manager, co-designer, co-producer, and co-manager of the living spaces. On the other hand, they must consider the role that urban design can play in designing health-based cities.

Particular reference is made to the capacity to generate new means of designing and organizing working spaces across scales, reconnecting and validating policies, projects, and plans that promote health and the well-being of city inhabitants.

In order to highlight and reflect on these relationships, this part of the book draws on some best practices. In New York and Toronto for North America and Medellín, Santiago de Chile, and Porto Alegre for South America, these themes hold notable importance in urban policies and, albeit from different points of view, aim to improve the quality of life in these cities.

An initial area of interest regards the effectiveness of involving and empowering local communities when affirming the policies, plans, and projects that promote the health of city inhabitants.

R. D'Onofrio and E. Trusiani, *Urban Planning for Healthy European Cities*, SpringerBriefs in Geography, https://doi.org/10.1007/978-3-319-71144-7_8

Although there is no certain proof, participation and empowerment are recognized as providing certain advantages (Zakus and Lysack 1998). These include: the increase of democracy, the mobilization of resources, the development of integrated, holistic, and sustainable actions, and the construction of decision-making processes, which tend to empower local communities (Morgan 2001; Smithies and Websterm 1998). In particular, citizen empowerment as defined by Laverack (2006)—"a process by which relatively powerless people work together to increase control over events that determine their lives and health"—would increase the ability of a community to identify and resolve problems. The acquisition of this capacity could lead to equitable and sustainable improvements in terms of health, especially for the most vulnerable swaths of the population and those least used to mobilization (Baker et al. 2005).

The use of a similar approach (UNICEF 1993) plays a rather important role in the attempt to unite urban planning with health. To favour this union, it is necessary that planning become "collaborative", thereby opening interaction with the local community to interpret the city and develop and identify the relevant choices. This occurs as in the interaction with the sectors and organizations operating there, such as universities, private organizations, local health services, etc. (Scotch and Parmanto 2006; Ashton 2009; Kazada et al. 2009). So that the interaction is really concrete, Flynn proposes establishing a broad structure for the community, encouraging participation, assessing needs, establishing priorities and strategic plans, soliciting political support, taking local action, and evaluating progress (Flynn 1996).

In this perspective, the search for planning models that place local communities at the forefront in empowering them to "produce" health and well-being becomes desired and necessary.

This is what was promoted, for example, in England already within the "National Planning Policy Framework" (NPPF 2012).[1] This law gives local authorities and their planning tools many responsibilities in terms of health, using "Neighbourhood Plans"[2] to establish paths to involve the population in order to promote the creation of safe, accessible, quality environments. In this regard, the NPPF requires preventive work to identify and resolve problems and to identify the requirements of places in terms of specific needs and quantitative and qualitative deficits or surpluses. In this research, the contribution from the local community is central and plays a primary role in assessing any impacts (HIA).

Assessing community needs and involving people in improving the quality of life in degraded neighbourhoods is a recurring theme in many English experiences. In Manchester, for example, the Manchester Neighbourhood Health Improvement Strategy (NHIS) of 2014, under the Manchester Health Department (MHD), proposed the creation of the Neighbourhood Health Framework. This strategy is based on six

[1] The National Planning Policy Framework (NPPF) of 2012 sets out the Government's planning policies for England and fixes the general guidelines within which local administrations should develop urban plans and assess the transformation proposals.

[2] The Neighbourhood Plans were introduced with the Localism Act in England in 2011. They enable communities to draft a plan for their area and make it possible for local communities to have a say in the development of the neighbourhood where they live.

domains to produce health: economic well-being, supportive living environments, access to appropriate care, healthy behaviours, social connectedness, and safety (City of Manchester 2014). The strategy foresees the involvement of the citizens through the activation of six Community Forums to gather their opinions regarding local needs, which have led to the development of the NHIS for neighbourhood-based investment in children and families. These Forums aim to share a vision of producing health through and with the population, with the aim of creating an orientation document to establish the collective impact of community-based health-improvement efforts.

The sharing of the principle that health involves everyone is reinforced by the presence of a leadership team that includes Manchester's Mayor, Police Chief, Superintendent, Public Health Director, residents, funders, and other community leaders to guide and facilitate the NHIS. From the sharing process produced, a series of priority recommendations were collected, many of which are closely related to the physical improvement of the places, including:

- creating leadership-training opportunities for youths and families with reference to the need to become involved and participate in efforts to improve the safety of the neighbourhood and the quality of life;
- establishing a coalition of key stakeholders to map the assets and analyze the differences in local resources. This is essential in supporting economic self-sufficiency on the individual and community levels.
- creating a coordinated, sustainable system of "Healthy Homes" composed of partnerships and multidisciplinary approaches that effectively and efficiently address questions about living conditions, which have a significant impact on health, safety, and the well-being of residents.

The recommendations/objectives are also accompanied by a search for possible financing to realize the interventions and a real business plan that promotes collective action through a deeper level of investment, such as the practices of collective impact and catalytic philanthropy, to support this large-scale, multi-sector initiative.

As an objective to be reached via collective involvement that is also transferred to city design, health has oriented the City of Belfast's policies for some years. As mentioned previously, in 2015, the city activated a process of community planning called the "Belfast Agenda", which will guide the city through 2035 (Belfast City Council 2017). To support the content of the Belfast Agenda, the "Belfast Conversation" was initiated. This is a series of workshops whose goal is to better understand the aspirations of residents, the community, and partner organizations regarding city development.[3]

Many themes, problems, and expectations in terms of health have emerged from the workshops. These include: the desire for a clean, healthy, and safe city, where everyone has access to good housing, quality green space, services, and structures that allow for a safe, active life. Some priorities have also emerged, such as: existing

---

[3] At: http://www.belfastcity.gov.uk/council/Communityplanning/TheBelfastConversation.aspx.

inequalities, risks for health, the transport emergency, accessibility and connectivity, and the search for green space (Pellegrino 2017).

In addition to the Belfast Conversations, on multiple occasions the city has promoted projects and plans attentive to the themes of health and well-being and even climate change that have seen significant involvement from the local population. This is the case, for example, with the Connswater Community Greenway (CCG). Beginning in 2006, the project was developed by the East Belfast Partnership in synergy with the Department of Agriculture and Rural Development (DARD), and received financing of £23.5 million from the Big Lottery's Living Landmarks programme.

The project to construct a 9-km-long linear park from Belfast Lough to the hills of Castlereagh was designed to improve the protection of about 1700 properties from flooding by expanding the culverts, realigning rivers, and constructing flood walls and embankments (CCG 2012). In addition to improving the physical environment, the objectives included the promotion of health and well-being of about 40,000 people living along the river by improving the river environment and green infrastructures (walking and biking paths, six tourist and historical interest routes, a civic square, etc.). Community involvement was strong already in the initial design phase of the project. Citizens took part in design laboratories to design the places and were actively involved in building a volunteer group to support the project.

One aspect of the role that local communities can play in promoting the health and well-being in cities regards co-managing the city spaces, and the City of Bologna is home to significant experiences in this direction. The presence of consciousness and responsibility in dealing with the management and design of public spaces was institutionalized in the *Regolamento dei Commons* [Commons Regulation] and *Patti di collaborazione* [Collaboration Pacts] (Labsus 2016). These are collaborative agreements that the administration signs with citizens, individuals, or associations, where the proponent is a co-planner and co-manager of a public space or public good in partnership with the City (Comune di Bologna 2016). This is the case of the "Corridoio ciclo-eco-ortivo—Quartiere San Donato-San Vitale" [Bike-eco-garden Corridor—San Donato-San Vitale Quarter] collaboration pact. Presented by a cooperative, the project regards the transformation of wild spaces into functional green spaces, such as social gardens for citizens and the production of vegetables and ancient grasses. City lands were transformed into garden spaces and productive agricultural fields. Another pact drafted by the City was the "Bologna Città Aperta: parco della Montagnola crocevia interculturale e multigenerazionale" pact [Bologna Open City: the Mntagnola Park, an intercultural and multi-generational crossroads]. This is a project to enhance the park promoted by the City in collaboration with several non-profit organizations. Cultural, leisure/recreational, artistic, handicraft, educational, and training activities are established, even with a historical and civil character. The scope includes: sharing and spreading the themes of diversity, culture, hospitality, solidarity, and anti-racism understood to be "common goods"; enhancing the liveability, pleasantness, and safety of the park; and favouring inter-generational and inter-cultural interaction through the development of activities aimed at precise age groups and targets.

These bottom-up projects, which encompass, spread, and interpret Bologna's mission of a "Healthy City", are situated within a tradition of collaboration and participation that has been consolidated over time and which is shown to be strongly rejuvenated and increasingly based on co-design. The projects and actions regarding the health and well-being of citizens do not always begin in a form that is united with the other sectors of city administration or coordinated via planning tools, as the assessor Rizzo Nervo reminds us in Appendix 1. However, they find communication, support, and collaboration precisely through the traditional mode of operation in the city, in synergy with all the areas of the city administration. The recognition of the administration's authority and the trust that citizens place in it is the key to success of this experimentation.

The experiences referred to in this Chapter show how the relationship wherein urban planning favourable to health works amplifies the need to draw on local and experiential knowledge of the urban environment reflected in the community.

In *Toward the Healthy City* (2009), Jason Corburn defines this knowledge as "local knowledge", which the community possesses and which can serve as valid aid to improve the production of living spaces. This is knowledge produced by concrete experience: the fruit of daily, emotional familiarity with the places, which at times can disagree with the knowledge/needs of administrators and the community of experts.

This interaction/conflict should give rise to the "co-production of knowledge", which ensures that technical problems are not detached from the social context and that a plurality of participants in the process are actively involved and contribute to making legitimate, shared decisions that make everyone responsible for "health". Operating in this way increases the possibility of finding effective solutions to problems pertaining to health and well-being, and improves citizens' trust and the credibility of public institutions.

Adopting an open approach to the conflict and local expertise also means not making the mistake of building a deterministically defined framework in the city's physical design, which would risk having to deal with abstract solutions to the problems of citizen health if these solutions, fruit of only "technical expertise", are not "tempered" by the history of the places and the people living there.

On the contrary, the "vision of places" allows for acknowledgement of how cities are complex systems that cannot be improved with isolated, permanent solutions, but which instead would present many adaptive, open solutions capable of being implemented and/or modified by the holders of different skills as needed over time. The "democratization" of knowledge on hot topics such as health and well-being in the city imposes not only a review of urban planners' traditional cognitive models, forcing them to modify protocols with which knowledge is built for the plan, but also to forge new tools for its creation. This is what emerges, for example, from the experience in Poznań,[4] which experimented with an interactive portal based on GIS technology to communicate with residents (appropriately trained and informed) in order to draft a shared version of the local development plan.

[4]See Appendix 2.

References

Ashton, J. R. (2009). From healthy towns 1843. *Public Health, 123,* 11–13.
Baker, S. M., Gentry, J. W., & Rittenburg, T. L. (2005). Building understanding of the domain of consumer vulnerability. *Journal of Macromarketing, 25*(2), 128–139.
Belfast City Council. (2017). *Your future city*. Ireland: The Belfast Agenda. http://www.belfastcity.gov.uk/council/Communityplanning/BelfastAgenda.aspx. Accessed June 6, 2017.
City of Manchester. (2014). *Manchester neighborhood health improvement strategy*. www.manchesternh.gov/health/NeighborhoodHealthImprovem. Accessed June 11, 2017.
Comune di Bologna. (2016). *2014–2016. Due anni di patti di collaborazione*. www.comune.bologna.it/.../due-anni-di-patti-di-collaborazione-il-report-e-10-storie-c. Accessed June 11, 2017.
Connswater community greenway. (2012). *Evaluation report-progress to march 2012*. http://www.connswatergreenway.co.uk/reports. Accessed June 11, 2017.
Corburn, J. (2009). *Toward the healthy city: People, place and the politics of planning*. Cambridge: MIT Press.
Flynn, B. C. (1996). Healthy cities: Toward worldwide health promotion. *Annual Reviews of Public Health, 17*, 299–310.
Gryboski, K., Yinger, N. V., Dios, R. H., Worley, H., & Fikree, F. F. (2006). Working with the community for improved health. *Health Bullettin, 3.*
Kazada, M. J., Beel, E. R., Villegas, D., Matinez, J. G., Patel, N., & Migala, W. (2009). Methodological complexities and the use of GIS in conducting a community-needs assessment of a large U.S. municipality. *Community health, 34*(3), 210–215.
Labsus. (2016). Rapporto Labsus 2016. Sull' Amministrazione condivisa dei Beni Comuni. www.Redattoresociale.it. Accessed June 11, 2016.
Laverack, G. (2006). Improving health outcomes through community empowerment: A review of the literature. *Journal of Health, Population and Nutrition, 24*, 113–120. http://www.bioline.org.br/hn. Accessed May 24, 2017.
Morgan, L. (2001). Community participation in health: perpetual allure, persistent challenge. *Health Policy and Planning, 16*, 221–230.
NPPF. (2012). *National Planning Policy Framework*. https://www.gov.uk/government/uploads/system/uploads/attachment_data/file/6077/2116950.pdf. Accessed February 14, 2018.
Pellegrino, P. (2017). *Belfast*. in D'Onofrio R., Trusiani E. (Eds.), Città, salute e benessere. Nuovi percorsi per l'urbanistica FAngeli Editore, Milan.
Scotch, M., & Parmanto, B. (2006). Development of SOVAT: A numerical-spatial decision support systems for community health assessment research. *Medical Informatics, 75,* 771–784.
Smithies, J., & Websterm, G. (1998). *Community involvement in health: From passive recipients to active participants*. Aldershot: Ashgate.
UNICEF. (1993). *Human development report 1993, Undp*. New York Oxford: Oxford University Press.
Zakus, J. D., & Lysack, C. L. (1998, March). Revisiting community participation. *Health Policy Plan, 13*(1), 1–12.

Chapter 9
Best Practices Around the World: Some Suggestions for European Cities

Abstract In addition to European experiences, which are often trapped within consolidated paths, in other parts of the world, some cities are opening up to welcoming experimental forms of small-scale bottom-up urban planning. Questions related to social sustainability are in many cases the bearers of innovation outside of conventional urban planning and design paths. The examples of North American cities rely a lot on experimentation through close involvement with civil society, which assumes responsibility for the city's living spaces, becoming a promoter of health-based initiatives, even from the economic point of view. Experimentation in the field is frequently accompanied by guidelines and tools charged with providing technical offices and designers with orientations rather than rigid rules. In addition, the impression is that it is the context that guides the interventions, avoiding simplistic, ineffective generalizations. With all the differences among the situations, health and urban planning are also themes addressed by South American cities. In many cases, they deal with approaches that are specifically health related and aimed at guaranteeing a minimum level of services, especially in the very diffuse areas of contemporary urban "informality" (*vilas*, *favelas*, etc.).

Keywords Urban resilience · Flexible urban planning · Bottom up urbanism
Performative urbanism · Resilient management · Participatory, inclusive planning

In an attempt to build experience and activate useful interactions to promote a new model of the health-based city, we have expanded our field of investigation outside Europe with a few specific cases. Either directly or indirectly, these experiences recall Urban Healthy Design and provide, in our opinion, some initial responses to the invitation in the recent Lancet article that "…urban design should be a globally relevant public health priority" (Sallis 2016).

The first experience regards New York City in the United States, which has been involved for some years on multiple fronts to promote the health and well-being of its inhabitants with policies, plans, and projects that regard the reorganization of urban spaces, the improvement of social and working conditions in the most rundown neighbourhoods, and information to promote better lifestyles.

Many of these initiatives, promoted by public and private institutions, use guides and manuals that orient design and encourage the virtuous, responsible behaviour

R. D'Onofrio and E. Trusiani, *Urban Planning for Healthy European Cities*, SpringerBriefs in Geography, https://doi.org/10.1007/978-3-319-71144-7_9

of the local population. The result is support for the much-cited need for a renewed alliance between city design and community empowerment in the name of health and well-being for all.

In 2006, the New York City Department of Health and Mental Hygiene promoted a close collaboration with the American Institute of Architects New York Chapter (AIANY). One product of this collaboration was "The Active Design Guidelines" (NYC 2010). This guide, which is accompanied by more detailed documents, offers architects and urban planners a manual of strategies to create healthier buildings, streets, and urban spaces based on the latest academic research and best practices in the sector. In particular, these strategies refer to:

- urban design strategies for creating neighbourhoods, streets, and outdoor spaces that encourage walking, bicycling, active transportation, and recreation;
- design strategies to promote active lifestyles through the placement and design of stairs, elevators, and indoor and outdoor spaces;
- discussion of synergies between active and sustainable design initiatives such as LEED[1] and PlaNYC.[2]

The Active Design Guidelines are a fruitful example of breaking down disciplinary silos in terms of health, since they were developed with the partnership of twelve city departments, university instructors, and the AIANY, to name just a few. These active design guidelines were incorporated into the design interventions established by the City with reference to the following main elements:

- developing and maintaining mixed land use in city neighbourhoods;
- improving access to transit and transit facilities;
- improving access to plazas, parks, open spaces, and recreational facilities, and designing these spaces to maximize their active use where appropriate;
- improving access to full-service grocery stores and fresh produce;
- designing accessible, pedestrian-friendly streets with high connectivity, traffic-calming features, landscaping, lighting, benches, and water fountains;
- facilitating bicycling as a means of recreation and transportation by developing continuous bicycle networks and incorporating infrastructure such as safe indoor and outdoor bicycle parking.

In 2013, this guide was supported with a detailed document, "Active Design. Shaping the sidewalk experience". Some of the principles contained therein seem to be inspired by good common sense: a well-designed sidewalk can favour pedestrian traffic, or well-lit spaces create safety. Other principles are not as expected, because they refer to psychological/intuitive principles that are implemented to increase the probability that residents choose the healthier option when using the city. The guide provides particular indications about what makes a route distinguishable, recognizable, and memorable and what makes it functional in relation to the size and

[1]LEED is a certification program that can be applied to any type of building (both residential and commercial). It concerns the entire lifecycle of the building, from its design through to construction.

[2]PlaNYC is a comprehensive sustainability plan for New York City.

proportion of human beings, the speed at which they walk, and the way in which they can see or perceive objects and activities. These indicators aim to provide tools to help politicians, designers, and citizens to support the choice of pedestrianization through the sharing of knowledge and projects (NYC 2013a). Another supporting document is the *Community Guide* (NYC 2013b). This guide is aimed directly at local communities and focuses on five aspects: active transport, active recreational activities, active buildings, green spaces and nature, and food and healthy beverages. Each section describes how the built environment can promote physical activity every day and how design can promote physical and mental health, social and economic vitality, and environmental sustainability. The guide orients citizens in intervening in neighbourhoods and assuming responsibility. The section related to transport, for example, develops ideas on how to make neighbourhoods more pedestrian friendly and provides information on how to undertake a series of initiatives, including: requesting a CityBench, pursuing the removal of graffiti, adopting a waste bin, or working with the local city council to request a slow zone. The approaches presented in this guide, which come from documented experiences and the object of interactive workshops, can be replicated by other communities in different quarters.

Another initiative launched by the City is the "NYC Plaza Program" (NYC 2016), directed by the Department of Transportation, which aims to transform underused street space into neighbourhood services through partnerships with local communities and non-profit organizations. The program, which began in 2008, has created 22 new public spaces for New Yorkers. Mostly temporary materials are used because the creation times are short and the rules for financing are more flexible, allowing community members to take advantage of them as quickly as possible.

The experience in New York is marked by numerous interventions promoted and financed by private foundations that regard both interventions on physical spaces in the most degraded areas and the creation of work, education, safety opportunities, etc.

The New York State Health Foundation (*NYSHealth*), a private, state-wide foundation, is dedicated to improving the health of all New Yorkers, with the "Healthy Neighborhoods Fund Initiative". This program is proposed to help New York State communities become healthier and more active places. The foundation has invested $2 million over two years to support six communities in the State of New York in their efforts to increase access to affordable healthy food; improving the safety of living spaces; and educating children and adults about healthy lifestyles.

Another foundation, the New York Community Trust (The Trust) has joined NYSHealth in a complementary initiative "The South Bronx Healthy and Livable Neighborhoods". The South Bronx is one of the poorest and least healthy communities in the United States. The program aims to improve the health and quality of life of the community by: expanding the availability of fresh, healthy, affordable food; creating safe streets, parks, and other public spaces conducive to physical activity; and improving job opportunities and income security for individuals and families, etc.

Another initiative from a non-profit community-development financial institution called Local Initiatives Support Corporation (LISC) instead aims to help community residents transform distressed neighbourhoods into healthy and sustainable communities. In particular, NYC's "Green and Healthy Neighborhoods Initiative" seeks to create healthier, more energy-efficient and environmentally sustainable housing and communities. This initiative, initially developed in response to rising energy and water costs, has grown over time into a comprehensive approach to developing sustainable communities by promoting a series of activities:

- providing opportunities to increase workforce skills in the green economy;
- improving residents' health with improved indoor air quality and non-toxic cleaning products;
- increasing access to healthy food;
- supporting comprehensive neighbourhood-based greening initiatives;
- renovating community gardens and planting street trees.

Finally, one of the latest initiatives promoted by the City was launched in September 2016 by Mayor De Blasio. This is a public/private partnership aimed at improving health in twelve neighbourhoods. "Building Healthy Communities" (BHC) is a multi-agency initiative concentrated on three main objectives: increasing opportunities for physical activity, increasing access to affordable healthy food, and improving the level of public safety. BHC involves $270 million of public investments in addition to $12 million of private financing. The twelve neighbourhoods affected by the project are East Harlem, Brownsville, Canarsie, Mott Haven, Hunts Point, Morrisania, Bedford-Stuyvesant, Harlem Central, Corona, Flushing, Mariners Harbour, and Stapleton.

The second experience worth noting regards the city of Toronto.

In 2011, the City began intense fact-finding activities through studies and reports on how cities shape the health of their residents. The first report, "Healthy Toronto By Design" (Toronto Public Health 2011) is organized as a sort of manifesto about health and well-being for the city and promotes a vision of the city to be realized through a strategic decision-making process and careful implementation of policies and projects that respect the needs and challenges of inhabitants. The report expresses the conviction that the challenge of health and well-being in the city is undertaken through design activities, intentional investment, and the provision of infrastructures, programs, and services centred on health. The report, which focuses on the role that local governments have in creating healthy cities, has been followed by others, such as "The Walkable City; Creating Healthy Built Environments: Highlights of Best Practices in Toronto", "Road to Health", "Enabling Healthier Neighbourhoods through Land Use Planning", "Toward Healthier Apartment Neighbourhoods", "Creating Healthy Built Environments: Highlights of Best Practices in Toronto", and "Green City: Why Nature Matters to Health—A Literature Review"[3], etc.

Tools have also been developed to assist decision makers in making decisions in the fields of health and well-being, with the "Health and Environment Enhanced Land Use Planning Tool" (Toronto Public Health and Urban Design for Health 2013).

[3]All of these documents are accessible at: http://www1.toronto.ca/wps/portal/contentonly?vgnextoid=e752105d4cff1410VgnVCM10000071d60f89RCRD.

This software program helps policy and decision-makers understand how different approaches to neighbourhood design might impact health-related outcomes such as physical activity levels, body weight, and greenhouse gas emissions. A technical report synthesizes information on the development of the tool and the results of pilot testing.

The City has also developed a guide—"Active City, Designing for Health" (Toronto Public Health et al. 2014)—focused on the city's physical built environment to create healthy places that encourage active living for all. The Guide outlines design principles to guide changes to neighbourhoods, streets, and buildings that allow people of all ages and abilities to incorporate physical activity into their daily routines without extra costs for physical exercise.

The guide outlines ten principles based on an active city and identifies good practices for each: "1. An Active City shapes the built environment to promote opportunities for active living; 2. An Active City has a diverse mix of land uses at the local scale; 3. An Active City has densities that support the provision of local services, retail, facilities and transit; 4. An Active City uses public transit to extend the range of active modes of transportation; 5. An Active City has safe routes and facilities for pedestrians and cyclists; 6. An Active City has networks which connect neighbourhood, to city-wide and region-wide routes; 7. An Active City has high quality urban and suburban spaces that invite and celebrate active living; 8. An Active City has opportunities for recreational activities and parks that are designed to provide for a range of physical activities; 9. An Active City has buildings and spaces that promote and enable physical activity; 10. An Active City recognizes that all residents should have opportunities to be active in their daily lives."

Finally, the guide establishes close links with some city policies and general and detailed urban plans promoted by the City, including: the City's Sustainable Planning Framework; Toronto's Zoning By-law; Toronto Official Plan's (OP); the Toronto Walking Strategy; The City's Traffic Calming Policy, etc. Other tools for planning and urban design address key aspects of built and public open-space design, and vibrant, attractive streets. Some of these are: the Urban Design Guidelines; the Streetscape Manual; the Bike Plan, Toronto's Parks Plan, etc.

The examples in North America rely a lot on experimentation through close involvement with civil society, which assumes responsibility for the city's living spaces, becoming a promoter of health-based initiatives, even from the economic point of view. Experimentation in the field is frequently accompanied by guidelines and tools charged with providing technical offices and designers with design orientations rather than rigid rules. In addition, the impression is that it is the context that guides the interventions, avoiding simplistic, ineffective generalizations. The approach taken in the United States and Canada serves as an interesting field of comparison for European cities to avoid falling into simple generalizations, especially where, for example, too much trust is placed in the HIA as a tool to validate plans and projects and not, instead, as a tool to support decision makers and their work. Great importance is placed on training technicians, politicians, and local communities to guarantee the creation of a more equitable city.

With all the differences among the situations, health and urban planning are themes also addressed by South American cities. In many cases they deal with approaches that are specifically health related and aimed at guaranteeing a minimum level of services, especially in those very diffuse areas of contemporary urban “informality” (*vilas*, *favelas*, etc.). Urban-planning actions that specifically interact with the determinants of health are not attended to. As also highlighted by de Leeuw and Simos (2017), the actions mostly deal with health policies to provide better services and important hospital centres, guarantee access to services for all, and/or interact with urban projects aimed at renewing and regenerating cities in terms of locating basic provisions and/or mobile health units in order to educate and raise awareness about these themes. Here, health and urban planning means primarily the right to the city, access to services, and the right to housing. If a city of the rich and a city of the poor exist, as Bernardo Secchi stated and as recalled by Patrizia Gabellini in the Preface, surely in many urban situations in the southern hemisphere this difference between the two cities is strongly accentuated and denoted by very strong contrasts resulting in numerous different variations. With regard to strategies implemented by national governments and local administrations, the intersection of health/urban planning themes occurs in some inspiring practices and in policies and specific actions that respond to the demand for resilience, intended especially in the sense of social sustainability, which, here more than elsewhere, becomes the unmentioned element of any form of a healthy and safe city.

In this sense, it is interesting to consider two documents that deal with urban policies on different scales, but which, in interscalar continuity with each other, offer directions and guidelines. These are the “International guidelines of urban and territorial planning. Toward a compendium of inspiring practices” by the United Nations Human Settlements Programme (UN-Habitat) for super-national policies, and documents relevant to the local urban-resilience strategies of three South-American cities: Medellín in Colombia, Santiago de Chile in Chile, and Porto Alegre in Brazil.

The first document, a report drafted by the United Nations Human Settlements Programme[4] (UN-Habitat) in April 2015, represents a meaningful point of reference for urban and territorial policies as well as a political, cultural, and operational anchor of notable importance on the prospect of research and design in contemporary cities.

The formulation and implementation of integrated policies, transformation strategies for renewal and planning, managing the environment, planning the city and compact, connected regions, participatory and inclusive planning. These are the five key lessons that the document assumes as the synthesis of inspirational practices. Within these, particular importance is placed on “participatory, inclusive planning”, which centres on the question of social sustainability in all phases of planning.

In 2003, Medellín adopted a holistic paradigm for urban planning that is identified with the practice of social urbanism. Urban transformations were directed at the most marginal and problematic areas of the city in order to make them safe and

[4]See: International guidelines of urban and territorial planning. Toward a compendium of inspiring practices, United Nations Human Settlements Programme (UN-Habitat); https://unhabitat.org/books/international-guidelines-on-urban-and-territorial-planning/. Accessed 22 May 2017.

guarantee accessible public places by connecting these zones to the city centre, using the environmental resources present as an element of connection and integration. The specific tool allowing these transformations was the Integral Urban Project (IUP), which established a series of projects for innovative development. It sets out a series of innovative development projects capable of improving public space and urban mobility of the inhabitants in these areas. Through the IUP, Medellín has managed to articulate its plan with specific actions to favour territorial dynamism, connect territories, integrate economic activities, promote social inclusion, and contribute to the creation of peaceful quarters. Inter-neighbourhood escalators, parks, libraries, and the Metrocable are now projects known throughout the world and are none other than the public face of a process to promote a new civic culture based on reinforcing participatory tools and empowerment.

The first objective of the resilience strategy, Equitable Medellín, aims to develop a more equitable and inclusive city and region with access to opportunities for all citizens. The actions implemented in the short and medium terms mostly revolve around the education/culture factor as a tool to discourage social inequality. With regard for the objective "Safe and Peaceful", the objective of reinforcing a more peaceful and socially cohesive society is set, promoting strategies to prevent crime and violence and engender a culture of legality.

The third objective, "Sustainable and Risk-prepared Medellín", is of particular interest. It aims to create a sustainable city prepared for risk through greater, conscious management of the territory and infrastructures via sustainable transport, environmental planning, and appropriate management of natural resources, as well as the quality of life of its citizens. Some of the pilot projects include "Strengthening of Community Risk Management", which aims to form a group selected from 102 local committees to manage risk in the neighbourhoods of Medellín. These committees are responsible for identifying and alerting communities about possible natural risks that can arise in order to adopt preventive measures to save human lives. The main arguments are centred on preventing catastrophes and on mitigation strategies during an event. It entails managing risks to the community, early-warning social networks, the social mapping of risk, and strategies for management during disasters.

In addition to prevention, knowledge, and information, we find meaningful pilot projects such as the one to reorganize informal settlements in hillside areas, which are at risk of landslides, and related education about how to prevent illegal housing solutions. These projects include the "Risk Knowledge and Reduction Program" (*Programa de Conocimiento y Reducción del Riesgo de Desastres*, CRRP), the Resilience Laboratory, "Risk Assurance and Transfer for Resilient Cities", the "Pilot Project for Preventative Resettlement of Land with Unmitigable Risk", the "Our River" project, and the strengthening of the Early-Warning Environmental System (*Sistema de Alerta Temprana*, SIATA). These projects address the theme of health from the viewpoint of securing the territory with respect to its endogenous characteristics and phenomena due to climate change.

Finally, the fourth objective of the resilience strategy, "Well-informed and engaged Medellín" means promoting a well-informed city through better access to information and data management in order to facilitate the analysis and transfer of knowledge for strategic planning and effective governance of the city. This is an important point because the pilot projects are the natural continuation of an innovative idea of community participation. In fact, the Agency for the Management and Integration of Information, the Communication Strategy "Medellin's Story Told Through the Transformation of the Region", the Think Tank for Creating Indicators, the Network of Social Observatories for Tracking Social Impact, and My Medellín were designed to monitor a participatory process that concludes with the Digital Strategy for Citizen Participation.

In the same way, Santiago de Chile organizes its resilience strategy, "Santiago Humano & Resiliente" [Humane and Resilient Santiago], on four principles that place humans at the centre of policies: *a human approach* that focuses the discussion on city inhabitants and prioritizes the quality of life in the city; *a participatory city* capable of integrating the different points of view and different desires of society representatives through effective tools for participation; *territorial intelligence*, which knows how to respond to the needs of the city and bases its strategies on concrete questions to answer; and *right to the city*, intended as access for all to the services and advantages that the city offers in order to promote equity, not only among the different social classes, but beyond the city borders to the numerous municipalities in the metropolitan region.

The resilience strategy is organized around these four principles and sets out six fundamental topics, 21 objectives, and 75 plans, programs, and specific projects to pursue the established goals. The six topics are: (a) urban mobility: prioritizing public transport and biking/walking over private transport by providing a reliable system of integrated, safe, sustainable, and smart mobility; (b) environment: promoting a city that grows and develops in harmony with the environment and uses natural resources responsibly and equitably, providing access for all inhabitants by creating green belts and quality urban parks; (c) safety: a city that promotes peaceful coexistence among its inhabitants and understands the multiple offenses and counters with coordinated, strategic, smart collaboration; (d) risk management: providing communities with knowledge and tools to be prepared for future disasters and therefore capable of reducing damage and the associated impacts; (e) economic development: promoting an equitable city that activates new opportunities within an ecosystem of innovation based on the circular economy, even on the regional scale; and (f) social equity: decreasing social gaps, inequality, and territorial fragmentation and guaranteeing access to services and opportunities for all.

The topics are made explicit in plans and projects, some already underway and others still at the proposal level. As already mentioned, these are projects that in many cases begin with sectoral governance and intersect the different components of managing and transforming the city and territory. The projects for mobility aimed at integrating and connecting parts of the city primarily respond to the demand for lower CO_2 emissions. They also answer the call for connections among different public transport services and biking/walking paths, intersecting parks and green areas distributed throughout the city, and making accessibility to places for relaxation, leisure, and community sports quick and safe. Within the overall vision, 75 projects have been established, including:

- Proyecto de modernización tecnológica de la gestión integrada de la movilidad urbana [Project to modernize the integrated management of urban mobility]
- Proyecto circunvalación intermedia para transporte público [Intermediate ring road for public transpot]
- Proyecto Nueva Alameda Providencia: remodelación del eje estructurante de la ciudad [New Alameda Measure: reorganizing the structural axis of the city]
- Plan Santiago Pedaleable [Bikeable Santiago Plan]
- Plan Santiago Camina [Santiago Walks Plan]
- Programa Regional de Cambio Climático [Regional Climate Change Program]
- Plan de Monitoreo Ambiental de Cuencas [Watershed Environmental Monitoring Plan]
- Plan de construcción de nuevos Parques Urbanos [Plan to Build New Urban Parks]
- Centro tecnológico de gestión de residuos [Technological Centre for Waste Management]
- Plan de descontaminación atmosférica Santiago Respira [Santiago Breathes Plan for Atmospheric Decontamination]
- Zona verde para el transporte [Green Zone for Transport]
- Programa Metropolitano de Consumo Responsible [Metropolitan Plan for Responsible Use]
- Programa de Protección de la Biodiversidad y el Ecosistema Hídrico [Program to Protect Biodiversity and the Water-based Ecosystem]
- Consejos comunales de seguridad [City Safety Recommendations]
- Plan de Recuperación sitios eriazos y espacios públicos deteriorados [Plan to Recover Vacant Sites and Deteriorated Public Spaces]
- Plan Red de Teleprotección Integrada [Plan for Integrated Safety Camera Network]
- Programa de participación ciudadana y coproducción de seguridad [Program for Citizen Participation and Safety Production]
- Plataforma de monitoreo de redes sociales en emergencias [Platform for Monitoring Social Networks in Emergencies]
- Centro integrado de la gestión de emergencias y desastres [Integrated Center for Emergency and Disaster Management]
- Programa de Riesgo Sísmico [Seismic Risk Program]
- Programa de Riesgo Hidrometeorológico [Rain and Weather Risk Program]
- Plan de Fortalecimiento Comunitario ante Multiamenaza en Pie-de-monte [Plan for Community Strengthening before Foothill Threats]
- Coordinadora regional de voluntariado ante emergencias [Regional coordination of emergency volunteers]
- Programa de educación para gestión de Riesgos [Risk-Management Education Program].

As only some of the proposals initiated, these demonstrate the 360°-involvement of the administration in making Santiago resilient, healthy, and safe. In this inter-sector framework, the contact between urban planning and health occurs around the theme of social sustainability. The involvement of citizens as co-producers of the city and their relative empowerment is fundamental and constitutes the key to reading all of the actions promoted.

Finally, beyond what is mentioned above, we mention the experience of Porto Alegre (Brazil), which, with the adoption of the strategic document *Desafio Porto Alegre Resiliente* [Resilient Porto Alegre Challenge] (2016), is predicted to reach the objective of a Resilient City in 2022. The document seems to be perfectly organized around the key lesson learnt from "participatory, inclusive planning" (UN-Habitat guidelines), and opens a new season in addressing the different economic junctions of the country and the current political crisis. This challenge is based on awareness and the mobilization and empowerment of all segments of civil society (citizens, communities, governments, businesses, social organizations, universities) in order to work together actively to resolve criticalities and render the city more resilient, harmonious, sustainable, and safe. The aim is to "help" the city best adapt to the impacts and pressures of today's world and to transform them into opportunities for growth. The concept of *urban resilience*, intended as the capacity of individuals, communities, or cities to survive, adapt, and grow, increasingly centres on the question of humans and citizens, and in this sense, is actively involved in all phases of the planning process. In this case as well, responsibility and empowerment are innovations with respect to the past and the transverseness of objectives and actions serves as the link between health and urban planning.

The health/urban planning intersection can be traced within the six large objectives of the strategy, which proposes a city with a (a) dynamic, innovative ecosystem, (b) culture of peace, (c) risk prevention, (d) quality mobility, (e) participatory budget, and (f) resilient management. With respect to the last area, it seems evident how the previous exerience with the management model becomes the starting point for the new objectives. The technical/management tool of the participatory budget is reworked and adapted so that it contributes to increasing the resilience index of the city and can promote a culture of resilience in all actions and interventions. Beyond the inter-sector projects involving the environment, mobility, and risk management, we cite two that can be defined as "a break" and "in continuity" with the past and which mark a turn in urban policies. Among the first, one can refer to the new role played by rural areas of the capital. These are no longer viewed as reserving building rights for the construction market, but, at least in the inspiring principles of the strategic document, represent opportunities, economic incubators for youths, biodiversity and nature, and investment for family agriculture, which triggers a circular process that is also able to address the question of urban poverty. Among ongoing projects, reference can be made first to the project to revitalize the Fourth District. This area is proposed as an upcoming centre between a healthy hub and creative activities developed from below. It is an experimental, innovative centre for rethinking the uses and means of activation, where the temporariness of some events has suggested including grass-roots actions and uses according to a principle of *mending*, going beyond the concept of regeneration. This is an innovative project where health and creativity are associated to transform a place and an entire urban area.

It seems clear how these South American cases also address a dual level of intervention—strategic and local—where the strategy guides local actions, which can occur even through forms of tactical urbanism, reinforcing the role of communities in the process of change. This can be a further element of interest for European cities, which are often blocked within consolidated paths and where innovation occurs in an organized, strategic way that is less open to welcoming experimental forms such as tactical urbanism, social urbanism, performative urbanism, etc. According to the most recent experimentation of small-scale bottom-up urban planning, questions related to social sustainability are in many cases the bearers of innovation ouside conventional urban planning and design paths.

References

de Leeuw, E., & Simos, J. (2017). *Healthy cities. the theory, policy, and practice of value-based urban planning*. Springer Publisher.

Sallis, J. F. (2016, May 28). Physical activity in relation to urban environments in 14 cities worldwide: A cross-sectional study. *The Lancet, 387*(10034), 2207–2217.

Websites

For New York City:

NYC. (2010). *Active design guidelines. Promoting physical activity and health design*. http://www1.nyc.gov/site/ddc/about/active-design.page. Accessed May 27, 2017.

NYC. (2013a). *Active design—Shaping the sidewalk experience*. In http://www1.nyc.gov/site/ddc/about/active-design.page. Accessed May 27, 2017.

NYC. (2013b). *Active design—Guide for community groups*. In https://www1.nyc.gov/…/active-design-community-guide.pdf. Accessed May 27, 2017.

NYC. (2016). *N*YC plaza program. Application guidelines 2016. In www.nyc.gov/html/dot/…/2015-plaza-program-guidelines.pdf. Accessed May 27, 2017.

For Toronto:

Toronto Public Health. (2011). *Healthy Toronto by design*. Toronto, Ontario. https://www1.toronto.ca/…toronto/toronto…health/healthy…/hea…. Accessed May 27, 2017.

Toronto Public Health and Urban Design for Health. (2013). *A health and environment-enhanced land use planning tool: Highlights*. https://www1.toronto.ca/…health/healthy…/clasp_tool_2012.pdf. Accessed May 27, 2017.

Toronto Public Health, City of Toronto Planning, City of Toronto Transportation Services and Gladki Planning Associates. (2014). *Active city: Designing for health*. https://www1.toronto.ca/City%20Of%20Toronto/Toronto%20P…. Accessed May 27, 2017.

For Medellín:

http://www.100resilientcities.org/strategies/city/medellin#/-_/. Accessed May 22, 2017.

For Santiago de Chile:

http://www.100resilientcities.org/strategies/city/santiago-de-chile#/-_/. Accesssed 22 May 2017.

http://www.municipalidaddesantiago.cl/. Accessed May 22, 2017.

For Porto Alegre:

http://www.100resilientcities.org/blog/entry/porto-alegre-releases-latin-americas-first-resilience-strategy-signs-10-res#/-_/. Accessed May 22, 2017.

http://www2.portoalegre.rs.gov.br/portal_pmpa_novo/http://www.100resilientcities.org/#/-_/. Accessed May 22, 2017.

Chapter 10
Urban Planning and Design Centred on Health Metrics

Abstract Recent years have been marked by numerous initiatives aimed at promoting experiments in the field, implementing the WHO's call in 2013 to move from the rhetoric of numerous policies aimed at promoting health and safety in the city to practical activities. All planning scales are asked to provide resilient design proposals capable of repositioning and reprogramming urban spaces in order to satisfy the needs of the community with respect to the potential impacts of urban transformations on the health and well-being of people. Reference to the context is fundamental, but it is also necessary to develop synergies among the different strategies and the different scales of the project. This includes a process of internal and external consultation in which the local community plays a fundamental role as the basic expression of the present and future social sustainability and resilience of the project. It is precisely in terms of resilience that some important cases of strategies, actions, and projects are found in Europe. Rotterdam in Holland and Copenhagen in Denmark are among the top examples.

Keywords Local community, neighborhood planning · Health metrics
Micro and macro urban design · Resilient plans and projects

10.1 Resilient Design Proposals for a Healthy City

The recent years have been marked by numerous initiatives aimed at promoting experiments in the field, implementing the WHO's call in 2013 to move from the rhetoric of numerous policies aimed at promoting health and safety in the city to practical actions (Rydin et al. 2012).

In 2015, for example, the English NHS together with Public Health England launched a new initiative to put health at the heart of new neighbourhoods and towns. The objective was to implement policies to construct 200,000 extra homes every year for the next five years and to refine the healthy city project to centre on the possibility of shaping places to radically improve residents' health and integrate health and care services. Ten cities were selected for a program that included "global expertise in spatial and urban design, national sponsorship, and increased local flexibilities"

R. D'Onofrio and E. Trusiani, *Urban Planning for Healthy European Cities*, SpringerBriefs in Geography, https://doi.org/10.1007/978-3-319-71144-7_10

(NHS 2015) in order to build new communities to support social cohesion, physical and mental well-being, walking, cycling, and sports. This was done via new ways of offering health and social services, new digital technologies to improve daily life, and new service infrastructures. The initiative, along with the Scottish design competition "Community Links Plus" (Sustrans 2016) and the Italian "Health City Think Tank" (HealthCity Think Tank 2016), aim to generate best practices and debates to address the themes of health and well-being, acting on the urban fabric and interpreting the urban past that plays such a role in pathologies of contemporary life.

An awareness lies in these programs and manifestos that urban design can connect data and policies to the experience of places with the possibility of transforming spatial practices and influencing the transformations that lead to positive changes for the life of people. This is a link to encourage, the connective fabric designers can use to contribute to macro and micro changes in European cities and cities around the world. The local approach recognizes and celebrates this moment: the places where we live contribute to the quality of our lives, in both good and bad ways.

All planning scales are asked to provide resilient design proposals capable of repositioning and reprogramming urban spaces in order to satisfy the needs of the community with respect to the potential impacts of the transformations and effects due to climate change on the health and well-being of people. Reference to the context is fundamental. One should try to understand, for example, relationships between: education level, family income, parks and open spaces, access to healthy food, obesity rates, etc.

Reading the interconnections among these different aspects is not enough, however, just as it is not enough to identify policies and strategies to face current social emergencies. It is necessary to determine the design's exploratory role in order to reposition and reprogram urban spaces with respect to the potential impacts of the transformations and effects of climate change on the health and well-being of city inhabitants. At the same time, it is necessary to develop synergies among the different strategies and the different scales of the project, even through a process of internal and external consultation in which the local community plays a fundamental role as the basic expression of the present and future social sustainability and resilience of the project. It is precisely in terms of resilience that some important cases of strategies, actions, and projects are found in Europe. Rotterdam in Holland and Copenhagen in Denmark are among the top examples.

10.2 Innovating the Approach to Redesign Existing Areas: Rotterdam and Copenhagen

In the delta city of Rotterdam, the theme of urban resilience as pertains above all to the safe city has become an urban policy of turning criticalities into resources, with particular reference to the city-water relationship. The Rotterdam Climate Initiative[1]

[1] See: http://www.rotterdamclimateinitiative.nl/uk/home.

can be considered the starting point for this experience. In turn, Rotterdam Climate Proof[2] is the tool that sets three basic objectives to be reached in the medium and long terms: Rotterdam–centre of excellence regarding water and climate change risk/management; Rotterdam—attractive city for new investments for the city and port; and Rotterdam—incubator of innovative applicative models/solutions to be exported elsewhere. The three objectives come together to delineate a strategy for a port city that is a model of sustainability, starting with the choice of field: coexistence with water and climate change and adopting the nature of a "sponge city". The Water Program, which is based on developing knowledge, short-term action plans, and the necessary professionalism, as well as sharing experiences (Mezzi and Pellizzaro 2016), modifies the consolidated approach and starts from the small scale. This micro experience has a high level of flexibility in the periphery, where squares based on water become the symbol of the new public space: green and blue, flexible and temporary use in agreement with climate conditions. These are created spaces visible to the population that are experienced and shared. Kleinpolderplein, Bellamyplein, and Benthemplein are the most recent projects.

The objective of security is the starting point. The basic infrastructure of canals and basins in the water system, collecting rainwater and mitigating run-off, and the reuse of water for irrigation purposes all trigger a design cycle where the formal success of redefining the open space is only the latest step in a complex rethinking of the city's basic infrastructure services. By changing the design approach and criteria, consolidated urban types (e.g., squares, gardens, parks, public spaces) no longer serve only for interaction, relaxation, and free time, but become fundamental elements for the safety of the city and its inhabitants.

Through design workshops, the involvement of residents and open-space users (Benthemplein, for example) contributed to examining the possible uses of the square to define its identity within the quarter and the acceptance of the project. Along with the floating neighbourhood under construction in the port area, these are significant pilot projects that are changing the face of Rotterdam, albeit always in ways consistent with its principal natural resource: water.

While the theme of water is a dominant factor, it should be recalled that the strategy of resilience is well structured. There are six areas of intervention that represent the main challenges for the city. First in order of priority, there is *social cohesion and instruction*, followed by *energy transition*, *climate adaptation*, *cyber security and use*, *infrastructure criticalities*, and *modification of urban governance*. The vision for a resilient Rotterdam is composed of transverse actions and initiatives following seven objectives: 1. Rotterdam: a balanced society; 2. Global port city built on clean, reliable energy; 3. Rotterdam Cyber Port City; 4. Climate-adaptive city; 5. Infrastructure ready for the twenty-first century; 6. Rotterdam network—really our city; and 7. Anchoring resilience in the city.

Each objective is accompanied by large actions that act as an economic flywheel as well as additional actions. The first are those that should guarantee a state of

[2]See: http://www.rotterdamclimateinitiative.nl/documents/2015-en-ouder/Documenten/ROTTERDAM%20CLIMat.%20PROOF%20ADAPTATION%20PROGRAMME%202013.pdf.

universal resilience, bouncing Rotterdam to the top of world cities, while supplementary actions should contribute with a smaller impact. This case also represents a twofold scenario between large and small actions.

The main priorities are social and human: building and reinforcing resilience on the individual and social levels. Starting with the assumption that knowledge, skills, education, health and well-being, and reciprocal understanding and respect are the central pillars of a balanced society, the administration supports and reinforces a certain number of current processes and initiatives to support the resilience of citizens and society through the Foresight Social program. Social sustainability, even in the case of Rotterdam, becomes the starting point for building all the strategies of resilience and healthy city, and education is the hinge around which change and growth revolve. This is related to the attempt to provide answers to the economic crisis, sudden social changes, and terrorist events that threaten daily life, placing people at the forefront of monitoring societal tensions and investing in social cohesion and resilience. The WE-Society Programme tries to answer to this scope, building an openness to diversity and reciprocal understanding among people as a given in maintaining social relationships among different groups present in the city. The social aspect is reinforced by the slogan "Qualified, healthy citizens in a balanced society", the first objective for an equitable society. Education to make today's young people competitive for the "next economy" together with the political document on public health (2016), which implements tools and actions for specific groups and problems, all seem to move in the direction of reinforcing social sustainability as the basis for a better quality of urban life.

The objectives and programs mentioned above are intersected by other projects that are more specifically oriented at transforming the physical space of the city and improving the quality of life. If Water-Sensitive Rotterdam is one of the crowning jewels, other projects have already begun. These include specific programs for the port (bioport), the perspective of the "next economy", the energy transition, and the Cyber-Resilience Platform, Cyber Resilience Desk, Cyber Resilience Co-op, and Cyber Resilience Officer to guarantee informational security. Others include specific interventions on the basic infrastructure such as burying infrastructure to make the city smart and easy to manage and reinforcing cooperation among all infrastructure managers for a common platform to share plans and knowledge, functions and interdependence among infrastructures located below and above ground.

Copenhagen, the European Green Capital of 2014, draws on the results of long-term policies for some aspects, and those that are feasible in the short term for others. In this respect, a key example is the brief period necessary to respond to the flood of 2011, which took the Danish capital by surprise, and to initiate and already partially realize the first projects to respond to climate change.

In terms of resilience and securing the city to improve the urban quality of life, Copenhagen has relied for some time on the Copenhagen Climate Plan[3], followed

[3]See: https://www.energycommunity.org/documents/copenhagen.pdf. Accessed 8 June 2017.

by the Climate Adaptation Plan[4]. The latter provides specific indications for interventions ranging from traditional open spaces—reinvented starting with the technological solutions adopted and from the flexible, temporary use for which they are destined—to updating the sewage and waste/disposal system of rain water in cases of extreme meteorological phenomena, to updating/rethinking ground floors and basements as areas for drainage. Ground floors would include additional areas to store rainwater, and basements would be used to create alert systems in case of rain in order to manage risk. These are actions and interventions aimed at preventing and managing climate change that lead to the relative transformation of the urban physical space. It is here where interest in the experience of the Danish capital lies. Beyond policies and strategies for resilience, actions initiated and realized on the micro level of planning make it clear how urban planning and urban design can affect the quality of life and health of inhabitants.

Policies enacted years ago to create a system of parks and areas for relaxation usable by all citizens and from any point in the city after a simple walk of about 15 min, and the ease of using bicycles for work/home and school/home commuting in the objective of a zero-emission city by 2025, for example, directly respond to the question of resilience. More indirectly, they respond to the request for daily movement, which is indicated as a basic requirement in preventing various twenty-first-century pathologies such as obesity, cardiovascular disease, and diabetes above all. This is a small but suitable example that ranges from citywide strategies and policies to interventions on the neighbourhood scale where, with citizen involvement, the best solutions for a city that responds to risks using liveable spaces are designed and discussed.

The theme of wastewater, which is also very pertinent to the city of Rotterdam, constitutes an important focus on which a radical intervention was operated as the result of a change in the strategic design approach: from risk to opportunity. The projects realized and initiated aim to alleviate pressure on sewer networks and at the same time to protect the city and its inhabitants; the techniques adopted refer to a cloudburst road, retention areas, and green roads, i.e., to the technological redefinition of basic (underground) infrastructures and the functional, spatial, and formal redefinition of surface areas. The application to individual neighbourhoods entails an overall renovation of public and private open space in the city and, by continuing the experimentation between neighbourhoods, creates functional and spatial connections between them, reinforcing their relationship to others in terms of community, sociality, and functionality. The project for the neighbourhood around Sankt Kjelds Plads is now an icon of change in Copenhagen and the visible synthesis among urban strategies and design. Aspects such as newly planted areas, the movement of earth to create two green dunes to increase the permeable surface area and reconfigure the pre-existing flat, monotonous space, the reduction of roadways and the creation of biking paths, and raising sidewalks to collect and drain excess water structurally and formally recount the transformation of public space and the entire neighbourhood.

---

[4]See: http://en.klimatilpasning.dk/media/568851/copenhagen_adaption_plan.pdf. Accessed 8 June 2017.

In this sense it is interesting to note how the Copenhagen Climate Plan explicitly dedicates a chapter to pocket parks and the role they play for health, well-being, environmental comfort, social aspects of the community, cohesion and sharing among residents and, not least, how they contribute to securing the quarter, representing "widespread green tiles" in the urbanized, impermeable fabric. It is not easy to find a call for a specific type of green in these strategic documents and directives and, in this case, a design action has perhaps still not been codified within the body of more traditional types of green areas. This highlights the multi-scalar nature of the document, the neighbourhood as a field of action, and the precise responsibility of urban planning in modifying the public space.

10.3 Between Macro and Micro: An Approach to Experimental Research

The experiences of Rotterdam and Copenhagen evince two design scales of reference for Healthy Cities: the strategic scale of urban policies, where the themes of the healthy city are found in a transverse manner, and the local neighbourhood scale where favourable conditions for experimentation seem to be concentrated. Based on the examples illustrated, the term *neighbourhood* seems to be returning to the centre of urban planning. It can be considered an experimental unit for the healthy city where a design core between the determinants of health and the quality of city public space can be sought without falling and/or seconding hygienist or welfare trends, as recommended in the Preface by Patrizia Gabellini, and even less by arriving at deterministic solutions by following pre-packaged paths.

Assuming the neighbourhood as the field for planning action means connecting to a recent past of modern urban planning. In fact, it has played a primary role starting with rationalist urban planning, which is viewed as an elementary cell in planning/designing the public and private city. The quarter constitutes the three-dimensional realization of the plan's provisions and represents the dimension of urban design, studying quantities, functions, and distributive rules. It was also the place where uniform populations from the social, demographic, or ethnic point of view were concentrated with recurring models of social interaction and precise systems of rules and local norms. For a long time, the quarter contributed to the growth of intense community life reinforced by the sharing of experiences, conditions, and by the development of a community understood to be a complex of elements among which the rooting to the places, identity and social recognition, relationships of reciprocity, relation, and solidarity stand out. The neighbourhood was a central theme in planning theory and research in the twentieth century, the reasons for which are technical, social, ethical, and moral. The public residential neighbourhood—rationalist and/or organic, characterized by a series of single-family homes or the territorial sign of building/city—has given form to theories of domestic and urban space in response to the essential needs of the population, always proposing new forms of community.

Today, the social/demographic reality is undergoing constant change and the idea of community is continuously brought into play by exogenous historical/social factors of a temporary and transitional nature. The quarter seems to express the basic potential to be able to activate experimentation in integrating the determinants of health and the network of mobility and access, the creation and diversification of widespread green areas, the activation of participatory forms of co-design, co-production, and co-responsibility, the innovation of public spaces starting with instances of securing areas, the redesign of technological networks in response to climate change, activating responsible participation processes, and the creation of consensus for shared responsibility. In other words, these criteria are at the basis of theories of the healthy city.

In more than one experience, reference is made to the neighbourhood as a local unit to apply urban-scale directives (London, Glasgow, Copenhagen, Rotterdam) or as a new protagonist in the process of co-responsibility for and co-production of the city (Bologna, Turin, Malmö, Copenhagen). On the other hand, assuming the neighbourhood as an autonomous object of study with the relative concepts of local society, neighbourhood, and a place where complexity is tamed or exalted due to its limited size—which makes it a controllable/monitorable tile in the urban mosaic—means it can serve as an important field of action precisely due to its location, which in many cases falls between decontextualization and reterritorialization. In this case, cohesion and social sustainability can become the potential on which to work for the transformation.

In a way somewhat similar to the past, the neighbourhood can express a new planning nature. In the 1920s and 1930s, the bases were formed for what after World War II would have developed on the large scale from national public building: urban planning models to guide the city's expansion and studies of the functions to rationalize/standardize building types and prefabricate building components. It is enough to mention the super-blocks in Vienna, transplants into the existing city fabric in Berlin and Frankfurt, the growth of satellite garden neighbourhoods and their separation with parks and agricultural areas, the planning of new neighbourhoods in Amsterdam, etc. In the 1920s, the New York Regional Plan defined the concept of *neighbourhood*, arriving at the *neighbourhood unit*, which has been refined in various ways in the guidelines of the examples presented. The sociological components are integrated with urban planning requests, including studies on the relationships among residence, collective structures, mobility, and social functions such as schools, parks, and shops.

It seems clear that there are affinities next to which the suggested road can be followed. For the rest, the first attempts at a healthy city, while not explicitly declared, can be traced to eco-neighbourhoods. These express the spatial results of growing attention towards environmentally friendly themes and sustainability via careful design for the efficient use of environmental resources, healthy and energy-efficient buildings, and land use compatible with ecological/social uses/sensitivities. These aspects fundamentally unite high architectural standards, energy savings, the drastic reduction of the use and possession of automobiles, the strengthening of public transport, and the incentivization of biking/walking. The Vauban quarter of Freiburg, the Zuidas quarter of Amsterdam, the Norra Djurgårdsstaden/Royal Seaport District of Stockholm, Lövholmen and Hammarby Sjöstad (always under the objective of Stockholm, a Fossil-Free City by 2050), and the BO01 neighbourhood in Malmö are only a few of many important examples. These are to be taken only as a partial reference since they constitute *ex novo* experimentation in many cases and not urban regeneration of the existing fabric; they also risk, in some cases, being situated as eco-ghettos for specific, exclusive social classes. For design criteria to be sharable, they should be widely applied and adapted to the local scale according to the situation, and within the existing city to rebalance situations of social inequality and improve lifestyles and the quality of life even and especially in the most fragile quarters, working towards social sustainability. As echoed many times in the text through the cases presented, as well as in the Preface by Patrizia Gabellini, it is perhaps the most important innovation and the basic principle for a healthy city.

With reference to the neighbourhood scale, what could the large categories of reference and objectives be to compose a grid of actions/options that respond to the health/urban-planning binomial, and what might constitute a useful reference in the design phase?

Some references in England show checklists organized around themes of healthy living, active movement, a healthy environment, and "lively" neighbourhoods. Community, housing, environment, and integrated transport with urban planning policies already provide a framework of reference within which to work and trace paths in line with the determinants of health. This is a less deterministic approach than the HIA, which can help designers and decision-makers understand the implications for health, local plans, and interventions for transformation. For the rest, the UK National Planning Policy Framework (March 2012) represents a collaborative approach between health and planning according to which local authorities accountable to city government should consider questions related to public health and collaborate with health organizations to understand the state of health and the needs of the local population in order to improve health and well-being. In this sense, the framework promotes a checklist to support those responsible for development proposals or planning, professionals in public and environmental health, forums of community groups, and local associations in order to contribute to involving the community and improving the proposed solutions. The checklist aims to combine the requirements and fundamental norms that influence health and well-being, providing support for decision-makers.

London's plan (July 2011) provides a meaningful framework for integrating health and territorial planning; it aims to improve health and address health inequalities. The checklist refers to the policies and standards of London's plan and the standards of quality and design, which are also inserted in complementary reports such as the "Code for Sustainable Homes", "Lifetime Homes and Lifetime Neighbourhoods", "Building for Life", and "Secured by Design" (2012). The checklist, which is divided into four themes, focuses on problems of health and well-being in relation to local strategies and strategies for health and well-being, such as those related to obesity and diseases linked to physical inactivity and inadequate food, air and noise pollution, road safety, social isolation, etc. (NHS 2014).

(1) On the other hand, Ann Forsyth, Emily Salomon, and Laura Smead, in their recent book *Creating Healthy Neighborhoods* (2017), provide eight principles around which to reason for integrated health and city planning:
(2) Importance: assess how health matters in this place;
(3) Balance: make healthier places by balancing physical changes with other interventions to appeal to different kinds of people;
(4) Vulnerability: plan and design for those with the most health vulnerabilities and fewest resources for making healthy choices;
(5) Layout: foster multiple dimensions of health through overall neighbourhood layout;
(6) Access: provide options for getting around and increasing geographic access;
(7) Connection: create opportunities for people to interact with each other in positive ways;
(8) Protection: reduce harmful exposures at a neighbourhood level through a combination of wider policies and regulations along with local actions;
(9) Implementation: coordinate diverse actions over time.

While based on an experience in the United States, this method is organized into three levels, always on the local scale: principles, proposed options, and action checklists. In Appendix B of the book, "Health Topics by Section", the proposed options and action checklists are related to the following health topics: air quality, climate-/heat-related illness, disasters, housing, mental health, noise, toxins, water quality, access to community resources, social capital, mobility/universal design, access to healthy food, physical activity, and safety. Table 10.1 illustrates the above-mentioned method, with reference to the "Principle 4 Layout".

This is naturally only an example with respect to the organization of the general principle, proposal, checklist, and health topic (in this case mental health, physical activity, air pollution and noise). Within each individual treatment, questions related to health/neighbourhoods are addressed in depth. This is the case, for example, with Principle 3—Vulnerability, where vulnerable populations and their health risk from negative place-related health outcomes is highlighted. The health topics already mentioned are related to vulnerable populations, i.e., low-income families, children, older adults, chronically ill people, women, ethnic minorities, city dwellers, rural populations, heavy labour workers, employees, and socially isolated people.

The same is true for Principle 6—Connection, where the positive and negative effects of social capital on health and wellness are highlighted. In the case of higher levels of social support, close relationships, and interpersonal trust, health and wellness impacts relate to better self-rated health (physically), better mental health, reduced mental disorders, reduced stress, and increased life satisfaction and happiness. Regarding the lack of social support, social isolation, low social capital, and loneliness, the health and wellness impacts are worse cardiac and all-cause mortality among patients diagnosed with coronary heart disease, increased risk of depression symptoms in general, related negatively to self-reported health, psychological and physiological stress. The authors likewise carefully describe health and wellness factors where social capital may have mixed or no effects. Therefore, for all the

Table 10.1 Organization of principle 4

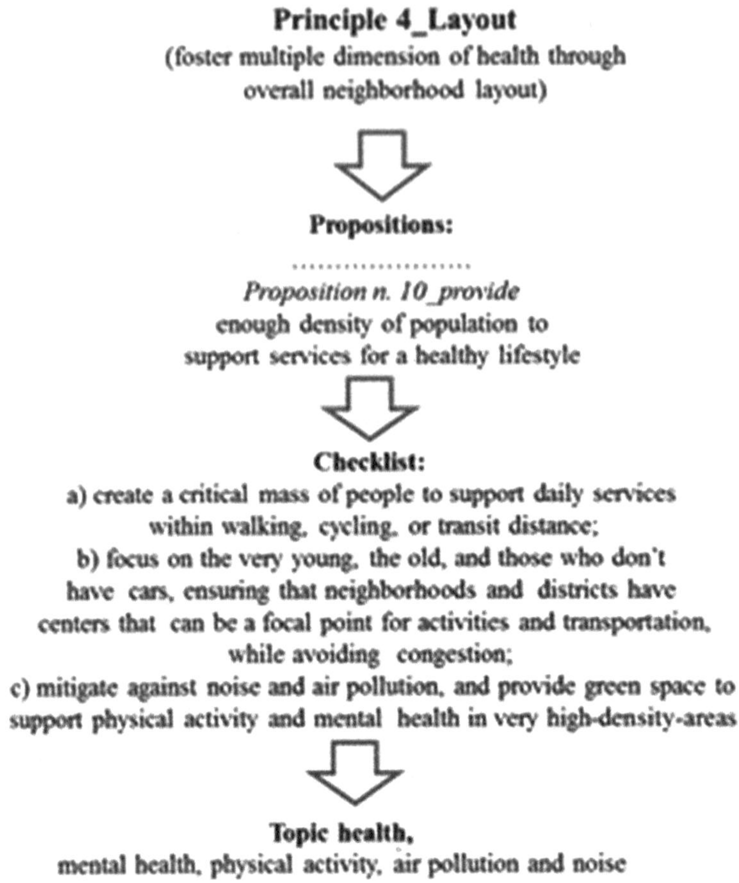

principles listed, there is a close correlation between health and the physical space used for living and interaction, i.e., the neighbourhood.

In the search for a renewed dimension in urban design open to the interaction with and contribution by local communities, other experiences suggest using guiding documents and principles that design proposals can refer to without the pretext of becoming binding rules, but precisely to interact with citizens and local operators.

In this respect, beyond the experiences in the United States and Canada referred to above, it is worth mentioning Scotland for Europe, which has developed the documents "Creating Places" (containing policies and directions for architecture and design), "Designing Street"[5] (containing principles and directions to design roadways, refurbish and maintain existing streets), and "Circulars" (regarding legislative and procedural devices). These specific documents are mentioned because they rep-

[5]See: http://www.scotland.gov.uk/Publications/2010/03/22120652/0.

resent a point of reference and connection between policies and actions/interventions, they initiate reflection, invite the actors involved into discussion, and, even going into the scale of their content, never assume the role of specialized technological and/or operational manuals for design.

The neighbourhood dimension, recourse to checklists that combine the fundamental requirements and standards that influence health and well-being with reference to plans and projects, reference to directions and guidelines that could be of assistance in formulating design proposals—in that they facilitate discussion among the different sectors of public administration, designers, local communities, and stakeholders—together can constitute a *modus operandi* for experimentation and refinement to construct cities oriented at the health and well-being of their inhabitants.

In proposing this path, new meanings are potentially acquired by two of the most debated aspects in contemporary urban planning: densification and the temporary nature of city uses. These directly and indirectly appear as basic requirements in almost all the experiences dealt with and in the theoretical references considered. In particular, the first appears as a sort of prerequisite for some recommended actions in terms of health, such as walking, socializing, sharing spaces, etc. The second appears as an opportunity to approximate the choice for quality design solutions over time in order to contribute to creating healthier and more equitable places and lifestyles.

In the existing city, the design application of the former leads to condoning demolition, a category of intervention that is used to open quality connective spaces between existing and new areas. These respond to requirements for connection, reduced auto use, biking and walking, quality public space, and multi-functional green spaces of every size to answer the demand for urban connectivity through green infrastructures (as well as climate change): actions that induce the change and better styles of life. In sum, densification seems to contribute to the realization of basic infrastructure of the healthy city as a prerequisite for recommended actions in terms of health, such as walking, socializing, sharing spaces, etc.

The second, the temporary nature of uses, directly introduces flexibility of use in the urban project in a double manner: "in expectation of" and "in use of". In the first case, the adoption of the design formula of the transition areas allows, within a limited time range, for spaces configured at low cost but with a high environmental return (in the case of a garden, park, etc.) usable by the inhabitants of the area undergoing transformation and the city residents. It serves as a sort of temporary incubator for quality, that is, an element for private operators to keep in mind as if it were a common green value in the moving real estate. In the second case, the squares in Rotterdam and Copenhagen are an emblematic postcard for the multi-faceted use of these spaces: from areas for play to squares for rain, from small arenas for shows and demonstrations to temporary ponds, from places for relaxing to places of transition and modification of the pertinent urban landscape. These spaces become modified and in turn modify the perception of their surroundings, becoming in a short time new places for social aggregation, landmarks in the temporary, changeable appearance, and cardinal points in renewed public space, as well as distinctive places for the communities that are attracted there.

References

Forsyth, A. Salomon, E., & Laura Smead, L. (2017). *Creating healthy neighborhoods*. Amer Planning Assn.

Health City Think Tank. (2016). *Manifesto La Salute nelle città: Bene Comune*. www.healthcitythinktank.org/Manifesto-la-salute-nelle-citta-bene-comune.pdf. Accessed June 11, 2017.

Mezzi, P., & Pellizzaro, P. (2016). *La città resiliente*. Milano: Altraeconomia.

NHS. (2014). *Healthy urban planning checklist*. http://tcpa.brix.fatbeehive.com/data/files/Health_and_planning/Health_Phase_2/Healthy_Urban_Planning_Checklist.pdf.

NHS. (2015). *The Forward view into action: Registering interest to join the healthy new towns programme*.

https://www.england.nhs.uk/wp-content/uploads/2015/07/healthy-new-towns-prospectus.pdf. Accessed June 11, 2017.

Rydin, Y., Bleahu, A., Davies, M., Davila, J. D., Friel, S., De Grandis, G., et al. (2012). Shaping cities for health: Complexity and the planning of urban environments in the 21st century. *The Lancet*, *379*(9831), 2079–2108. https://doi.org/10.1016/s0140-6736(12)60435-8.

Sustrans. (2016). *Community links plus*.

http://www.sustrans.org.uk/scotland/communities/community-links-plus-design-competition. Accessed June 11, 2017.

MIX
Papier | Fördert gute Waldnutzung
FSC® C083411

Kolibri 360 GmbH
Ferdinand-Jühlke-Straße 7
99095 Erfurt, Deutschland
produktsicherheit@kolibri360.de